The Environment in Question

This book provides an introduction to the key environmental debates. By addressing specific global problems and placing them within an ethical context, the collection provides the reader with both theoretical and practical understanding of environmental issues. The contributors are internationally known figures drawn from a range of different disciplines, including geography, psychology, social policy and philosophy. The contributions range from those tackling individual concrete issues (such as nuclear waste and the threat to the rain forest) to those addressing matters of policy, principle and attitude (such as our obligations to future generations and the nature of technological risk). Emphasis lies not only on scientific facts, but on ethical perspectives – principles of trust, co-operation, far-sightedness, respect and concern for the future.

The Environment in Question is designed as a text for students of philosophy, environmental science, environmental education, ecology and teacher education. It can be used as a self-contained, inter-disciplinary course book or in conjunction with relevant material. In addition, this collection of previously unpublished essays will interest professionals in each field, as well as the interested layperson concerned about this planet's future. The substantial cross-section of concerns and approaches will provide all readers with the necessary background and insights to develop an awareness of the problems and to enter into the environmental debate.

The Environment in Question

Ethics and global issues

Edited by
David E. Cooper and Joy A. Palmer

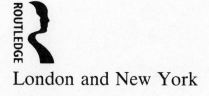

London and New York

First published 1992
by Routledge
11 New Fetter Lane, London EC4P 4EE

Simultaneously published in the USA and Canada
by Routledge
a division of Routledge, Chapman and Hall, Inc.
29 West 35th Street, New York, NY 10001

Phototypeset in 10/12 Times by Intype, London
Printed and bound in Great Britain by
T J Press (Padstow) Ltd, Padstow, Cornwall

British Library Cataloguing in Publication Data
The Environment in question : ethics and global issues.
 1. Environment. Ethics
 I. Cooper, David E. (David Edwards) II. Palmer, Joy
 179.1

Library of Congress Cataloging in Publication Data
The Environment in question : ethics and global issues / edited by
 David E. Cooper and Joy A. Palmer.
  p. cm.
 Includes bibliographical references and index.
 1. Human ecology—Moral and ethical aspects. I. Cooper, David
 E. II. Palmer, Joy A.
 GF80.E57 1992
 179′.1—dc20   91–16823

ISBN 0 415 04967 9
ISBN 0 415 04968 7 pbk

Contents

Figures

Tables

Notes on contributors

Dr Robert Allison is Lecturer in Engineering Sedimentology at the University College, London University.

Professor David E. Cooper is Professor of Philosophy at the University of Durham.

Dr Barry Gower is Senior Lecturer in Philosophy at the University of Durham.

Professor Sir Frederick Holliday is formerly Vice-Chancellor and Warden at the University of Durham, now Chairman of the Joint Nature Conservation Committee.

Professor C. A. Hooker is Professor of Philosophy at the University of Newcastle, New South Wales.

Mrs Mary Midgley was formerly Senior Lecturer in Philosophy at the University of Newcastle upon Tyne.

Mr Philip Neal is General Secretary of the National Association for Environmental Education.

Dr Joy A. Palmer is Director of the BA (Ed.) Degree Course and lectures in education and environmental issues at the University of Durham.

Dr Robert Prosser is Head of the Department of Geography, Newman College, University of Birmingham.

Professor Holmes Rolston III is Professor of Philosophy at Colorado State University.

Professor Mark Sagoff is Director of the Institute for Philosophy and Public Policy at the University of Maryland.

Dr Vandana Shiva is an environmental scientist at the Research Foundation for Science and Ecology, Dehra Dun, India.

Mr Stephen Sterling is a consultant in environmental education.

Dr Rosemary Stevenson is Senior Lecturer, Department of Psychology, University of Durham.

Dr Jennifer Trusted is part-time Lecturer in Philosophy, University of Exeter Department of Continuing and Adult Education, and a tutor with the Open University.

Editors' preface

The proposal for this book arose out of a BA (Ed.) course on environmental issues taught at the University of Durham, in which several of the book's contributors have been involved. It became apparent to us that, despite the recent explosion in the literature, there existed no volume of essays suitable as a main text for such a course, one to which students could be referred in connection with most of the topics they would be examining.

A suitable volume, we felt, should aim to be reasonably comprehensive in two respects. First, it should contain contributions by authors from a variety of disciplines which bear upon environmental issues – geography, psychology and ethics, for example. Second, the chapters should range from ones tackling specific, concrete issues (such as nuclear waste and the threats to tropical forests) to ones which address broad matters of policy, principle and attitude (such as technological risk and obligations to future generations). It would, of course, be impossible to include in a single book, which is to be both affordable and manageable in size, contributions from *all* the relevant disciplines and upon *all* the topics of current environmental debate. We hope, however, that the present volume offers students a substantial cross-section of approaches and concerns.

While the book is intended for use as a text in wide-ranging environmental courses, this is not its sole purpose. All of the essays are published for the first time, and several of them offer new analyses of the issues and make controversial proposals for their resolution. These essays will, therefore, be of interest to professional academics, whether in the sciences or the humanities, and to those with responsibility for environmental policy. More important, perhaps, the essays – all of which avoid technical discussion as far as possible – will be of interest to a public whose concern for environmental issues has very rapidly expanded in recent years.

The contributions are, to repeat, highly varied in topic, scope and approach, and there is no one, obvious way in which to divide them into sections. The reader will find, however, that the first nine chapters of the book focus either upon specific moral questions or upon particular issues of concern to environmental science. The remaining chapters address more general themes and concepts (such as sustainable development and technological risk) which permeate thinking about a wide range of specific matters, including those discussed by authors in the earlier chapters.

1 What do we owe future generations?

Barry S. Gower

I

People have, and have always had, ambivalent feelings about the future. In a rapidly changing world, where what has happened is often an unreliable guide to what will happen, we are apt to agree with Francis Bacon's observation that 'men must pursue things which are just in present, and leave the future to the divine Providence'. What has not yet occurred cannot be a legitimate object of our concern and care, because for all we know it may not occur at all. On the other hand, we cannot help but be aware that what happens in the future is less a matter of 'Providence' than it is of what we do in the present. We can, that is, affect the future even though we cannot know it. This, no doubt, is why we spend so much of our time in formulating plans, not only for our own personal future but also for the circumstances in which we as well as our descendants will live. When things go wrong with our lives, or with the environment in which we live them, few of us are sufficiently philosophical to blame fate for what has happened. We blame, instead, the 'planners' for their lack of forethought, or for their insensitivities, or for their sheer ignorance of what does happen.

This ambivalence has consequences for our thinking about our obligations to people who, barring global disasters, will live in the future – to future generations. Given that we have, now, unprecedented power to affect the lives of these people in very many ways, we may well think that the interests of generations not yet born should weigh just as heavily with us as the interests of people alive now. For it is only the fact of someone's being a person, and not the further fact that she is a person *now* rather than at some date in the future, which should have any bearing on her moral standing.[1] Yet we do not seem to act accordingly, especially when it is lives that will be lived in the distant future that are affected. When we undertake an analysis of the advantages and disadvantages of some social policy bearing on resource depletion or ecological exploitation, we tend to discount to some degree the interests of future people, so that the foreseeable consequences of what we do matter less and less the more remote in time they become. Thus, in evaluating a waste-disposal policy, or a decision to neglect some part of our cultural heritage, we are likely to consider its modest benefits to us and our children as outweighing its greater costs to subsequent generations. It is as though the moral value of a policy to future generations were expressible in monetary terms and would therefore be steadily eroded as a result of inflation.[2]

It would be facile to take these tensions as evidence of inconsistencies. The life, the vigour and the importance of moral thought depend upon tensions like these, and no one need be surprised to learn that questions about our obligations to future generations, when pressed far enough, lead to problems and paradoxes. It is quite true that these problems have a theoretical nature which makes them seem academic and unworldly – an impression which is reinforced by the fact that the examples used as an aid to discussion are schematic and simplified. Nevertheless, they are just as important as the more practical social, economic and political problems that face us, though not for the same reasons. There are, of course, good reasons why we should seek acceptable solutions to our practical problems. The quality of our everyday lives depends on such solutions. But in the case of the philosophical problems explored in this chapter, there is a different reason for concern. It derives from the fact that our capacity to reflect upon, plan for and care about the future is part of what makes us human beings. So if we are to have a proper understanding of ourselves as human beings, it is important that we explore the limits of this capacity. One way to do this is to address the question of our obligations to future generations.

II

In our more benign moments at least, most of us think that we *do* have obligations to future people. We *owe* them something and *should* care about their interests. We believe, for example, that if 'greenhouse' gases produce significant global warming, then for the sake of our descendants steps should be taken to limit the severity of its effects by curbing the over-production of such gases. We ourselves, it is said, are already beginning to observe the consequences of global warming on our climate and ecology, and it will take time, measured perhaps in generations, for these to be reversed. Similarly, we think that because the effects of projected levels of marine pollution are long-lasting and will be felt most severely by people several generations in the future, we should refrain from using the sea for the disposal of sewage sludge, hazardous waste and radioactive contaminants. Even if we cease using the oceans as industrial sinks forthwith, many years will elapse before the sea's regenerative powers can counteract present levels of pollution in some areas. And on land, the hazards associated with nuclear power installations are, we think, at least as important for future generations as they are for us. Chernobyl has shown us what devastating effects nuclear accidents can have, and has made us more conscious than ever of our obligation to do whatever we can to prevent such accidents from occurring again.

Yet despite this view that something should be done to protect the interests of future people, there is wide disagreement about how we should meet the challenges presented by global warming, marine pollution, nuclear power and many other environmental issues. In part, these disagreements arise because there are a great many uncertainties about *how* future people will be affected by what we do, uncertainties which increase as we try to envisage the needs of people living in the more remote future.[3] We know so little about what human life will be like in a thousand years' time, if indeed there will then be any life that is recognizably human, that it is hard to see how we could judge the effects upon it of what we do now. Even within a more limited time scale, the evaluation of alternative policies is a complex task, involving not just technical knowledge, but also expertise in weighing uncertainties. We would need estimates of many social factors, such as economic growth rates, population growth and distribution, energy use and conservation and agricultural production. Needless to say, predictions of these kinds of factors which concern more than a very few years would need to be treated with the greatest caution.

But it is not only these kinds of uncertainties which must be tackled. For even if we knew what the effects of our policies would be, and how they would be felt, there remains a neglected but legitimate doubt about how far, if at all, the interests of people who do not yet exist are to be included in our moral deliberations. Should we, as an American poet provocatively asked, act

> . . . as though there were a tie
> And obligation to posterity.
> We get them, bear them, breed and nurse:
> What has posterity done for us?[4]

This doubt, its source, and the difficulty of finding a satisfactory resolution for it take us to the heart of our topic.

III

Moral questions arise when what we do affects the lives of other people, and one kind of effect which we must consider is that which involves the satisfaction, or its thwarting, of a need.[5] Suppose, for example, you learn that a stranger is suffering from a form of leukaemia for which he needs a bone-marrow transplant if he is to survive. You happen to be the only known person who can donate the right type of bone marrow. By agreeing to help meet the leukaemia victim's need you do something that is, to that extent, morally good. You act as a 'Good Samaritan' and the reason that will commonly be offered for saying that such an act is praiseworthy is that it promotes the well-being of the person in need, and any action which promotes well-being is, to the extent that it does so, morally good. On the other hand, if without good reason you ignore the need, then you become liable to moral criticism on the grounds that you are callous, unfeeling, inhumane.

Appeals to humanity and to the principle of beneficence which this virtue expresses can be powerful, especially when eloquently presented. Newspaper advertisements for Third World charities commonly include photographs of undernourished children in the expectation that humanitarian feelings will be aroused. And it is true that when we see a clear need and are sure that our contribution will enable it to be met, many of us are willing to respond. The trouble is, of course, that the needs of undernourished children, though real, are all too easily forgotten, and that even when they are not it is sometimes hard to convince us that we can meet them. This perhaps is one reason why the humanitarian case, although so simple and strong, has not proved particularly successful in eradicating poverty, disease and deprivation in many parts of the world.

A Good Samaritan attitude towards the plight of future generations has also been less than prominent. Faced with the complexities of modern life, we sometimes find it hard enough to manage the *immediate* effects of our actions, let alone the long-term consequences. Concern for the future is – or so it can seem – a luxury that we lack the leisure to indulge. No doubt our apathy is partly due to the difficulty of sustaining an appeal to humanitarian sentiments when we cannot see or photograph the hardships suffered by children of future generations. Nevertheless, those hardships are the consequences of our policies and are often preventable at relatively minor cost to us. So the case for action on behalf of future people would appear to be a simple one. And yet we consistently find that our response, when there is one, is inadequate. It would be easy to say that this failure on our part is just another illustration of the general truth that we do not live up to even the most rudimentary of moral standards. Sometimes this is no doubt true, but in other cases we should consider whether there might not be some feature of our thinking about future generations which would account for our inaction.

One way of trying to explain our reluctance to embrace thoroughgoing humanitarian aims for future people depends on the observation that each of us can affect the lives of future people – apart from immediate descendants – only to a very small extent. One person is willing to incur minor inconvenience in order to save energy; another is not willing to forgo even the slightest convenience and squanders it. In neither case are there perceptible effects on others, especially when we project those effects into the future. Again, I have a choice between voting for policies concerning waste disposal which are cheap for us but highly dangerous for posterity, or for policies which entail expense for us but greater safety for posterity. In a hundred years' time, I reflect, my vote will make no difference. But if the consequences of what I do for the sake of posterity are negligible so far as it is concerned, it would seem that it is only the more immediate and perceptible effects on us that need to be considered in reaching a judgement as to the moral worth of what I do. You act with compassion when the effects of what you do go some way towards meeting a need, say that of future people. But if this need will be just as great whatever you do, it might seem that your compassion would be better directed towards other needs – say, those of your contemporaries and immediate descendants. Individually, none of us can benefit or harm to any significant extent the lives of future people, so there can be no obligation arising from the principle of beneficence to exercise the virtue of humanity on their behalf.

This is an argument that we can and should resist. True, an action can be right or wrong because of its effects, but we should consider that action and its effects in a context where the actions of others are relevant. Your lone attempt to alleviate hardships suffered by future people, or by any other large number of people, may make negligible difference, but that same attempt when coupled with the efforts of others to the same end could well make a significant difference. It is because your attempt and its effects can be part of a larger-scale project that it can have moral worth. There are, then, good reasons for claiming that an individual action can be right or wrong on account of its effects, even though those effects are not noticeable. If this seems paradoxical, that is because, in accordance with the Good Samaritan image, we are apt to focus on individual acts of beneficence or maleficence, and to overlook the benefits and harms which can only come about as a result of co-operative action. For our thinking about our responsibilities to future generations, it is particularly important that we recognize how significant are the consequences of what we, together, do so far as future generations are concerned, even though the effect of what each individual does is so trivial. And in judging the morality of our actions by their results, it is the significant effects of what we do *together* for such matters as environmental pollution and resource depletion, rather than the insignificant effects of what we do individually, that are to be counted.[6]

There is, though, another and better reason for believing that a principle which enjoins us to promote well-being is not sufficient for a proper understanding of our responsibilities towards future generations. This reason is based on the claim that a principle of beneficence is liable to the charge that it will condone injustice. For example, many people believe that it would be quite wrong to deprive further the poor of our own generation for the sake of advantages for future generations, even if that would increase the total amount of well-being. The end may be worthy, but in this case the means to its achievement are not. Issues like this are at the centre of current debate about the need in developing countries to adopt those practices widely used in developed countries which are now thought to be environmentally damaging. It cannot be right to use the needs of posterity to justify our failure to respond to the comparable needs of the present. Even in those cases where the means available for securing advantages for posterity are not considered to be morally objectionable, we might wonder whether we should use them if those future people affected will in any case enjoy a higher level of well-being than us. Suppose, for example, we have a choice between conserving or depleting a limited resource – say, fossil fuels, or a safe area for waste disposal – and are confident that, whichever policy we choose, future people whose lives will be affected by our choice will lead lives more worth living than our own. In such circumstances we can and, arguably, should choose to deplete even though, compared with conservation, this means a large loss of benefit for future people, and will mean less overall well-being for everyone affected.

To an extent, the idea that ends justify means could itself be called harmful in that it threatens the autonomy of people. But it is even more important to observe that when, as in the above examples, it increases inequalities it gives unacceptable conclusions. There are bound to be inequalities between generations, as there are between contemporaries, if only because those living later are likely to know more about how natural resources can be used to meet human needs. So when considering exchanges between generations, we must take inequalities into consideration before drawing conclusions about how much saving one generation should undertake for the sake of its descendants.[7]

The humanitarian case for extending our care and concern to future people is, then, less persuasive than it might be because the principle which underlies it is subject to some important qualifications. The biblical Good Samaritan sacrifices only his own convenience, but our dealings with future people require the active co-operation of many of our contemporaries, and their agreement that current claims – like the claim to equal treatment – as well as current needs, should be given up to secure benefits for future people. In short, though humanitarian aims are entirely praiseworthy, some ways of achieving them are morally unacceptable, and the basis of the judgement that they *are* unacceptable is clearly something that must be taken into account in our thinking about our obligations to future generations. It may be that some of us will respond to the imagined needs of future people in much the same way, and with much the same motives, that others respond to the photographed needs of some presently existing people who suffer from poverty, disease and deprivation. But we lack, as yet, a way of qualifying the general principle which justifies these responses. Qualifications are necessary because resources directed towards future people, like resources directed towards the Third World, cannot then be made available for use elsewhere, and there can be no guarantee that their use elsewhere will not produce superior benefits.

IV

We turn, then, to that important part of our behaviour towards each other which has to do, not so much with questions about how, if at all, needs can be met, as with questions about what is fair and just. Thus, although compassion may lead me to sell all my possessions to raise money for a group of starving refugees in Africa, I anticipate that those nearer to home who have a *claim* on me for their support will do their best to persuade me that my obligations to them have priority. They will point out that there are some things that I must not do – this being one of them – even though the well-being of others would be thereby improved. Often, the language of 'rights' is used to express these obligations. In a general sense, respect for rights does promote the well-being of all, but in practice there are many circumstances in which we have to decide whether a person's rights should or should not be overridden in order that well-being can be promoted.

In the case of contemporaries, it is possible to understand our obligation to treat each other fairly as being based on implicit *agreements* that we, as independent rational agents acting out of self-interested motives, make with each other. There are, as it were, unwritten rules which govern our conduct towards each other. These rules are not imposed on us, but made by us in our own interests. Each of us benefits from the continued willingness of others to abide by the rules, and anyone who violates these rules, even though much good comes from the violation, acts unjustly. In this way, questions about what is fair are questions not about subjective attitudes or feelings, or about what will increase the sum of human well-being, but rather about what it is rational for co-operating agents to do.

But although it makes some sense to speak of contemporaries entering into such agreements with each other, it makes no sense to speak of contractual obligations, however implicit, governing the conduct of present people towards future people. For agreements must benefit each of the parties involved, and even though it is true that we can benefit future people, there is nothing that future people, apart from our immediate descendants, can do for us in return for the favour. How, then, can we say that obligations to act fairly towards future people have any role in constraining what we may do to enhance well-being? We certainly act selfishly and with a callous disregard when we ignore the interests of future generations; but do we then also act unfairly? It would seem not, unless we can identify another basis for the conclusion that we do.

Consider Mabel, who has a good life – as do all her friends and acquaintances. She is well fed, healthy, and lacks little that she wants.

And compare her with Melissa, who, like her friends and acquaintances, is often hungry, frequently ill, and lacks most of what she wants. To a significant extent, the predicament that Melissa and her friends find themselves in is the result of the indulgent lifestyle enjoyed by Mabel and her friends. If Mabel and Melissa were neighbours, fellow citizens or even just people living in different countries at the same time, it could with justification be said that Mabel is behaving unfairly in not sharing her fortunate circumstances with Melissa. It could even be said that in such circumstances Mabel *owes* Melissa a share of the resources she uses to sustain her well-being. For we can understand this obligation as arising from a tacit agreement they made with each other to benefit from each other's co-operation, and Mabel *is* benefiting from Melissa's co-operation, or at least from her tolerance of the situation.

But suppose Mabel is my grandmother and Melissa my granddaughter. It would no doubt have been a kindness if Mabel had arranged to share her fortunate circumstances with her great-great-granddaughter; but did she *owe* her something? It would seem not. The past is fixed and no one can now or in the future change it, so Melissa cannot do anything for Mabel in exchange for whatever Mabel could have done to alleviate Melissa's misery. There is no agreement, no possibility of mutual co-operation, and therefore, so it seems, nothing Mabel *owes* to Melissa.

There has, though, always been an alternative tradition of thought according to which an agreement or contract cannot be said to determine what is just or unjust, but must itself be determined as just or unjust. Independent objective standards of justice – natural justice as it is called – enable us to judge the fairness or otherwise of whatever contracts, real or implicit, we make with each other; they enable us, that is, to judge whether what counts as just 'by convention' is really just. If this is so, then questions about justice between generations can be asked; and if we know what is required by natural justice then we will be able to answer these questions.

Many believe that equal consideration for all is fundamental to our moral tradition and is required by natural justice.[8] So, inasmuch as the contracts made by self-interested and rational people are bound to ignore the interests of those who, for one reason or another, have little or nothing to contribute, they are unjust. The moral importance of a person's interests have nothing to do with whether it is possible for that person to bargain with us. Everyone, regardless of their circumstances, has a right to equal consideration.

In some contexts it does indeed seem that we endorse an equality principle by conceding that the needs of future people should be given as much weight as the needs of contemporaries. If, for example, we confine highly toxic waste in a container which we know will leak, causing the deaths of hundreds of people, then it makes no difference to our degree of culpability if the leak and its disastrous consequences occur next year, in a hundred years, or even in a thousand years.[9] Perhaps we believe that in a thousand years this waste will be less of a threat because, say, an immunity will have developed, or an antidote will have been discovered. But this means only that the needs of people living then should be considered differently, not that they should be disregarded. Equal consideration does not always mean equal treatment.

In general, our pollution of the air, the sea and the land affects us as well as our descendants. Anything we are able to do to abate pollution levels will benefit people living now as well as everyone in the future. Moreover, there is no incompatibility between our enjoyment of fresh air, clean water and uncontaminated land, and that of our descendants. There is, then, no serious moral difficulty about how the costs and benefits of pollution control are to be shared between generations. This is not to say that complacency is justified. There are very difficult political, social and economic problems that we have to face up to if the lives of future people are not to be made intolerable by pollution.

But there are other contexts where the right to equal consideration seems to set an unrealistically high standard. We do not always consider remote descendants equally with immediate descendants, and an equality principle which tells us that we should is less than compelling. If, for example, the right of equal consideration were to imply, as it might, that everyone is to have an *equal* share of some finite resource then it may be that no one would have enough to make any difference to their well-being. To some extent this could be overcome by the members of one generation co-operating with each other so as to ensure that some benefit was derived from their combined share, but the problem of how members of different generations could or should co-operate remains. A more realistic view of what justice requires would be one which implies, not that we share our resources with the *whole* of posterity, but that we make *some* savings for its sake. The moral problem associated with resource depletion is whether this requirement can be justified and made less vague.

We can, I believe, make some progress with this problem by using the idea that the *expected* benefit or harm of a policy concerning finite resources is subject to a 'discount rate', not on account of time as such, but because the probabilities of some benefits and harms become smaller as we project them further into the future. For example, if the current generation curbs its use of fossil fuels then they will be available for the next generation of users who, we can be sure, will have a need for them which is comparable to our own. By contrast, saving fossil fuels for a remote generation, say that living a thousand years from now, would make much less sense. We cannot assume that that generation's need will be at all comparable to our own; nor can we assume that our savings will be passed on by the intervening generations. The probability that our savings will benefit that generation is, therefore, very small, and so also is the value to that generation of our saving, i.e. its expected benefit.

With the aid of this concept, the idea of equal consideration can be given a different interpretation. It is not so much people, or their interests, which should be considered equally, but rather the expected benefits and harms they derive from our actions. Suppose, for example, that a man buries a fortune on an island, worth £A, and believes, correctly, that there is only a small chance, B, of its being discovered by the island's sole inhabitant. So far as this islander is concerned, the 'value' of the man's action – its expected benefit – is the product of the fortune's worth and its probability, i.e. £AB, which will of course be less than £A. The same benefactor gives the single inhabitant of another island the smaller sum of money, £AB. Have the two islanders been given equal consideration by the benefactor? In an important sense, they have, because £AB is a fair price to pay for the privilege of taking either islander's place. Similarly, suppose Mabel bequeaths a fortune – £A – to the eldest among her great-great-grandchildren with blue eyes and fair hair, and believes, correctly, that there is only a small chance, B, that such a descendant will exist. The expected benefit of Mabel's bequest is £AB, considerably less than £A. She could, alternatively, give the smaller sum £AB to a contemporary relative whose need, we can assume, is just as great as that which Mabel's bequest would satisfy. Is Mabel giving as much consideration to her blue-eyed and fair-haired descendant – if there is one – as she is to her contemporary relative? It would seem so, because £AB is a fair price to pay for the interest of either beneficiary in Mabel's choice.

There is, then, some justification for giving priority to the needs of our own and our children's generations.[10] The expected benefit of satsifying a particular need of these generations will be greater than the expected benefit of satisfying that same need of future generations, and the proposed interpretation of the equality principle requires us to take into account these differences. Furthermore, resources saved for some are, in practice, resources lost to others, and this interpretation provides a way of balancing the one against the other. With a policy of conservation, the present cost of resources saved for the future represents a determinate loss of expected benefit, whereas the uncertain value of future benefits of those savings means that they must be discounted before being put in the balance. With a policy of depletion, the present benefits of resources used represent a gain in expected benefit, whereas the future cost of resources no longer available must, again, be discounted. This is not to say, of course, that expected benefits and harms for future people cannot outweigh expected harms and benefits for present people. It expresses, rather, the idea that injustice can occur when benefits for our contemporaries and our children are sacrificed for the sake of greater, but only probable, benefits for future generations.

Our knowledge concerning anything but the near future is, and must remain, very limited. Certainly, we can be sure that even remote descendants will have some of the same needs as we do. This means that the value of some of our savings will decline little, if at all, over time. The basic biological needs for food to eat, air to breathe, water to drink and space to live will remain, no matter how far into the future we look. On the other hand, the value to remote descendants of our endeavour to save, say, fossil fuels must be less – perhaps much less – than its value to our children.

We are not, then, obliged to sacrifice everything for the sake of posterity. But we are required by justice to sacrifice something, for there are limits to what we can legitimately do to increase well-being. So long as we recognize that justice limits but does not prohibit our humanitarian attempts to enhance well-being, there is no conflict between the principle that demands equal consideration for all, and our intuitive feeling that the immediate needs of our contemporaries and their children should feature more prominently in our moral deliberations than the conjectural needs of remote descendants.

V

Until fairly recently little careful thought had been given to questions about the nature and extent of our responsibilities to future generations. The natural concern expected of parents for their children and grand-children was sufficient acknowledgement of such responsibilities. But as we have acquired ever more power to control the future, these questions have become more pressing. We cannot any longer suppose that it is just our immediate descendants who will incur the consequences of our choices. Some current environmental policies are, we are told, imperilling the long-term ability of the planet to sustain human life, and unless we change these policies people of the further future will be obliged to pay for our profligacy with their well-being. Vigorous political creeds purport to express the moral implications of this fact. But the glib answers to practical moral questions that they provide are hardly an adequate substitute for the recognition that a moral problem, however practical, derives its meaning from a sophisticated context of reflection. We have inherited an enormously rich stock of ideas about how moral issues are to be examined, tested and resolved. In using these ideas, we cannot expect clear-cut solutions, for ideas have complex histories and subtle inter-relationships. Often, there is a dialectic, or opposition, between one set of ideas and another, and this can lead to important insights even if the dialectic remains unresolved.

No wonder, then, that when we use our legacy of moral ideas in an attempt to understand new questions about our obligations to future people, we find no easy solutions but, instead, perplexing problems.

Whether the lives of future people will be better as a result of our philosophical study of these questions remains to be seen. It is certain, though, that our lives can be made better, for in examining the bearing of our moral tradition upon our thinking about future generations, we will be exploring an important aspect of the culture which determines our capacities and limitations.

NOTES

1 See e.g. Sidgwick (1877: 383).
2 For criticism of the view that the moral importance of an action declines over time, see Parfit (1984: 480–6). See also Moore (1903: 152–4) and Smart and Williams (1973: 33–4).
3 See Kavka (1981).
4 John Trumbull, quoted by G. Hardin in Partridge (1981: 221).
5 So for example, Ross writes of 'duties of beneficence' which 'rest on the mere fact that there are other beings in the world whose condition we can make better in respect of virtue, of intelligence, or of pleasure' (Ross 1930: 21).
6 For further discussion, see Parfit (1984). Parfit's book also contains extended treatment of the complex problem of how a principle of beneficence should be expressed, given that our decisions about environmental problems will affect not only *how* people in the future will live, but also *who* will live.
7 See Rawls (1972: 284–93).
8 See Glover (1984: 140–52).
9 See Routley and Routley (1981).
10 For a different way of arriving at this conclusion, see Passmore (1980: 87–100).

2 The problem of absolute poverty: What are our moral obligations to the destitute?

Jennifer Trusted

What obligations do we have to others? If, far away, people are starving, are we morally obliged to give all that we have (short of starving ourselves and our dependants) to save or to attempt to save them?

Some recent writers have maintained that not to give is tantamout to murder. They hold that there is no morally relevant difference between killing and letting die. James Rachels argues that what appears to be an intuitive belief that letting die is not as bad as deliberate and unjustified killing arises because such a belief suits us in terms of costs and benefits. We have much to gain from a strict prohibition on unjustified killing and it costs us very little to conform. By contrast we have little to gain as recipients of charity and it would cost us a great deal to be donors (Rachels 1986: 148). Peter Singer concedes that selfish indifference, though blameworthy, is not as culpable as active malevolence, and he admits that to cut down our standard of living to bare essentials does 'require a degree of moral heroism utterly different from what is required by mere avoidance of killing'. He also concedes that those in affluent countries cannot be held to be directly responsible for suffering elsewhere. However, he concludes that these and other differences are not sufficient to undermine his view 'that there is no intrinsic difference between killing and allowing to die'. He argues that we are responsible for our fellows, though he tempers his criticism by suggesting that while 'not aiding the poor is not to be condemned as murdering them; it could, however, be on a par with killing someone as a result of reckless driving, which is serious enough' (Singer 1979: 163–8).

Singer considers that if we are to be strictly moral we should donate to the point where we live in real austerity, but he appreciates that this sacrifice is unlikely to be made and that to urge such self-denial as a moral duty is likely to be not only fruitless but also counter-productive. Hence, since his aim is to help those in need, he holds that it is morally acceptable to seek to persuade people to sacrifice less, in the hope that there will be a wide response to a limited appeal. Any suggested limit must be arbitrary but he thinks that there is something to be said for suggesting 10 per cent of income, 'more than a token donation, yet not so high as to be beyond all but saints'. Clearly the richer citizens of affluent countries can spare more both in absolute and relative terms than can the poorer (perhaps from 20 to 90 per cent of their incomes should be expected) but the moral argument applies to all who have means above those necessary for subsistence; a teacher cannot absolve herself of her duty to give by saying that merchant bankers must first make their sacrifice.

Singer and Rachels both affirm our moral obligation to all who are in need regardless of where they live. Indeed Rachels thinks to give his point greater force by saying that the duty to feed a starving child in Africa is as great as the duty to feed a starving child sitting in the room in which we ourselves are about to eat.

I intend to argue against this view. I am not concerned with objections based on the practical difficulties in making sure that funds donated actually benefit those for whom they are intended. I agree with Singer that though such objections can be ethically significant if it is certain (or almost certain) that gifts will not be received, objections based on the mere possibility of wastage and partial loss on account of inefficient distribution and mismanagement (or embezzlement) are morally irrelevant. They are usually introduced as an excuse for indifference. Nor am I concerned with allotting blame for misery and destitution; nor whether it is the case, as Nielsen (1984) says, that the policies of western capitalist countries are responsible for mass starvation; or, for that matter, whether the policies of the eastern block have led to impoverishment in the Third World. Many complex issues are involved:

> The actual pattern of colonial violations of economic rights is complex and obscure. In the heart of darkness everything is murky. Many former colonies were economically backward when colonized; some colonial administrators did a good deal to modernize and develop the infrastructure of their colonies; it is always uncertain what the present would have been had the past not been colonial.
>
> (O'Neill 1986: 110)

I doubt if any individual anywhere can be said to have made no indirect contribution to human suffering but we are not here concerned with relating duty to alleviate distress with responsibility for that distress. We are concerned with each person's moral obligation to help others just because they need help. To what extent are we our brothers' keeper? In the original context Cain was referring to his brother with whom he had grown up, whereas most references today are to be taken in a metaphorical sense. It might be better if we did love each other as siblings but the plain fact is that we do not. We do not feel as deeply for the interests of strangers and we do not believe we have the same obligations to them as we have for those near to us.

James Fishkin argues, and I shall develop this argument later, that an assumption of general obligations to all would be overwhelming and that

> For this reason, all persons of at least ordinary capacities must be vulnerable in theory to a sufficient number of acts required by *positive* general obligations as to overwhelm the limits that, under normal conditions, had appeared perfectly appropriate. . . . Because the obligations are general, everyone of at least ordinary capacities is, in theory, subject to them.
>
> (Fishkin 1982: 50)

We have no general moral obligation of beneficence, although I do not suggest that we should be indifferent to suffering in distant countries and indeed I shall argue that we do have a duty to support activities aimed at promoting a fairer distribution of resources. But I contend that our charitable duties are and must be much more limited than Singer and Rachels suppose; beneficence is never a matter of duty, and arguments for beneficence based on appeals to moral logic are misguided. This is not to say that there is no place for beneficence, or that the love of one's fellows, charity in the biblical sense, is irrelevant. Indeed it is the basis of ethics and I am inclined to take the Pauline view that it is the highest virtue.

> Though I speak with the tongues of men and of angels, and have not charity, I am become as sounding brass or tinkling cymbal.
>
> And though I have the gift of prophecy, and understand all mysteries, and all knowledge; and though I have all faith, so that I could remove mountains, and have not charity, I am nothing.
>
> And though I bestow all my goods to feed the poor, and though I give my body to be burned, and have not charity, it profiteth me nothing.
>
> Charity suffereth long, and is kind; charity envieth not; charity vaunteth not itself, and is not puffed up. . . .
>
> Rejoiceth not in iniquity but rejoiceth in truth.
>
> (*Corinthians* (1), ch. 13, 1–6)

There are no arguments here; the assertions appeal to ethical feelings rather than to reason.

At the last, all ethical judgements arise from emotion but there does seem to be a place for moral logic when we consider the sphere of duty. Hence I propose a distinction between duty and benevolence in that duties entail moral obligations and can, at least in the first instance, be demonstrated by moral logic, whereas benevolence is not obligatory and should be evoked by appeals to the heart rather than to the head. The distinction depends on accepting that our general moral obligation to others is an obligation to avoid harming them, that is, to avoid causing pain and to avoid being unjust. Moreover I would extend this general obligation to certain species of animals in that I believe that there is a prima-facie obligation to avoid causing pain and distress in respect of all sentient creatures. This is not to say that it is always immoral to kill animals nor even to condemn eating animal flesh, but to say that very many species are proper objects of moral concern.

If, following Rachels and Singer, we accept, in addition to these negative obligations, general positive obligations for all sentient creatures then we must acknowledge that we have strong obligations to the destitute and the victims of calamity everywhere. Moreover if physical location is irrelevant, so that it is irrational to say that distance reduces obligation, it is also irrational to base one's obligation on the nature of the suffering or the age of those suffering and perhaps even their species. It follows that if we have a moral obligation to try to prevent death from starvation we must also have a moral obligation to try to prevent death from infirmity or disease. Hence we must have a moral obligation to help those who need heart transplants, kidney machines and vaccines. Moreover if we are not to show speciesism we have a moral obligation to make the same sacrifices in order to save animals: starving herds as well as individual animals. In addition, since not only life itself but quality of life is important we must surely be under strong moral obligation to help people with physical disabilities and the mentally handicapped. We have, if we accept the moralists' argument, a moral obligation to give up everything (short of that required to maintain our own lives and the lives of our dependants) to help all our fellow creatures at all times and in all places. Our obligations would be truly overwhelming.

The fact that many sentient beings need help does not completely undermine the moralists' argument (in one sense it strengthens it) but it does show that they are too parochial in that the kind of obligation that tends to monopolize their attention is only one of many. Even if their thesis of general positive obligation is entertained, it is clear that, though everyone who can must make great sacrifices, no particular duty of charity, such as feeding the starving in distant countries, is categorical. Thus though my (and Fishkin's) argument does not show that it is irrational to acknowledge a moral obligation to those starving in faraway countries it does show that there is no absolute duty to help and it would not be morally irrational to decide to devote our inevitably limited resources near to home. However, I concede that this is not to offer an argument to counter the moralist thesis that ideally we ought to give away everything that we have, short of maintaining life at subsistence level, to those in need.

There are three reasons for objecting to such a sacrifice. First, if exhortations to sacrifice everything short of that required for subsistence are successful they are likely to be self-defeating. Second, it is not wrong to favour those who are near and dear to us; and third, pleasure is an important part of all lives, including our own, and the quality of all our lives, including our own, has some place in moral assessments. Finally,

though not an argument against obligatory beneficence, it is worth pointing out that great stress on ameliorating conditions through 'hand-outs'
may lead to the underrating or ignoring of political actions. Charity can
but mitigate suffering; political reform may end or reduce it.

Let us consider the first point, that if they are successful exhortations
for sacrifice are likely to be self-defeating. Most people concerned with
this book, writing it or reading it, will be very fortunately placed in
relation to the great majority of their fellow human beings. They will be
well fed and will almost certainly live in considerable comfort; many are
likely to be able to take holidays, to travel and to own a car. Indeed
probably the majority could live in fair comfort with substantially less
than their present income; and of course even then they would be truly
affluent compared to those in the Third World. Admittedly most of us
will have markedly less than the wealthiest in western society but, as
already indicated, the moralists would argue that this does not remove
or even reduce our personal moral obligation to make sacrifices; nor is
the moral obligation of those who have considerably less than the average
(but who are still above subsistence level) reduced. If the moralists'
argument is accepted, almost everyone in the affluent countries, and
even some who live in the Third World, are under a direct and personal
obligation to give to those who are in distress. The moralists who urge
the obligation to make sacrifices must hold that all should make that
sacrifice. It is not to the moral point to say that there will only be a
small response, so that the western world will continue as an affluent
consumer society.

Now in the affluent countries a relatively small number of people work
to produce food and to tend the sick; the rest are concerned with
producing the goods and services that make life comfortable, interesting
and relatively luxurious. We live in a market economy, and though, in
a primitive market economy, people will work hard for the necessities
of life, in a prosperous market economy mere existence is taken for
granted. People work in order to acquire the 'extras'; the desire to
acquire comforts and luxuries is a major incentive for working. I do not
say that it is the only incentive; those fortunate enough to have interesting work will also work because they enjoy the work itself and take
pleasure in what they achieve. But even for them the prospect of enjoying
life beyond subsistence level plays its part. Would the desire to help
unfortunates in distant countries be such a strong incentive for extra
work? It is likely that, if it were possible to establish a market economy
with no consumer goods and services beyond those necessary to maintain
life, people would not apply themselves so diligently. They would prefer
to cultivate their gardens. Not a bad thing perhaps. It is highly probable
that we should all be happier and much less stressed if we opted out of
the rat race for 'goodies' and lived more simply, but there would be
little to spare for the starving, almost certainly less than today. A market

economy of this nature, even if it could be established, would not help the poor.

How about a controlled economy? Perhaps it might be possible to persuade a philosopher king that benevolence was morally obligatory and that it was immoral to enjoy creature comforts when others in the world were starving. In that case we can envisage a controlled economy in which only essential goods and services are produced, in which each individual is rationed and in which all surplus is sent to those in need. This kind of economy would be similar (though even more austere) to those found until recently in the Soviet bloc and we have seen how inefficient they were. In such economies production is very low partly because people have to waste time queuing for necessities (and therefore have less time to work) and partly because, and for the same reasons operating in a market economy, there is little incentive for working beyond a level needed to maintain existence.[1] Hence, even if moral exhortations were successful and such an economy were established, those exhortations would be self-defeating.

Singer and Rachels would, I suspect, claim that this is not to the point and that what they are saying is that all of us ought to work just as hard to provide necessities for others as we do to provide comforts and luxuries for ourselves and our dependants; morally we should make no claim whatsoever on the goods and services our society produces, beyond that needed for our bare subsistence. Moralists must, of course, have moral ideals but I suggest that it is disingenuous to advocate action that has no chance of success. If the moralists ignore this and implicitly rely on a partial response they are not seeking the fulfilment of the general moral obligation that they claim is demanded by moral logic.

A way of avoiding this difficulty would be to suggest donations of about 10 per cent of income, as advised by Singer. But we have to bear in mind that for Singer this was a moral compromise – half a loaf (or rather 10 per cent) being better than no bread – for he is clear that ideally there should be total sacrifice. However, we need to bear in mind that he suggested 10 per cent just because he had come to the conclusion that it was not realistic to press for total sacrifice. It is interesting that elsewhere in his book he implies the morality of a less than heroic self-denial in that he states that it might be necessary to 'keep up appearances' in order to earn more money and thus to give more to help others. He says:

> If my present position earns me, say, £10,000 a year, but requires me to spend £1,000 a year on dressing respectably and maintaining a car, I cannot save more people by giving away the car and clothes if that will mean taking a job which, although it does not involve me in these expenses, earns me only £5,000.

(Singer 1979: 163)

If everyone behaved in the way Singer advocates as an ideal he would be earning very much less in any case, and in addition there would be no need to keep up appearances. More importantly this very concession to appearances is a sign of the weakness of his thesis of moral obligation. As Onora O'Neill says, for Singer everyone has a moral duty to give: 'The important point in Singer's position is not that it leaves too little room for anything but beneficence, but that it makes it entirely clear that beneficence is a matter of obligation' (O'Neill 1986: 57).

It must be stressed that this, my first objection, is a practical not a moral one and it may be contended that I am merely making a pragmatic case for the respectability of selfishness. Indeed Singer's concession to the need to keep up appearances can also be taken as a concession to selfishness – it is pleasant to keep up appearances! However this would be unfair to me and very unfair to him. My point is that the moralists' position cannot be rationally justified because the self-abnegation, which they set up as a moral ideal, would be self-defeating. However just because one is not prepared to make or to argue for a total sacrifice it does not follow that one condones total selfishness. My arguments are not designed to show that the concept of benevolence is spurious, they are designed to show that the concept of a general positive moral obligation of beneficence is suspect.

My second objection was that it is not wrong to favour those who are near and dear to us. As I write, an indefinitely large number of sentient creatures are suffering, many very deeply. We do not need to be profound pessimists at least to entertain the thought that throughout all ages there has been more misery than joy and more pain than pleasure, so that a utilitarian calculus would show the balance of unhappiness far outweighing happiness. Perhaps not, but whatever view we take about the general happiness of all sentient creatures there can be little doubt that the world would be a happier place if *Homo sapiens* had never evolved or if our species became extinct. There is plenty of good evidence to support the belief that annihilation of the human species would be in the general interest of other sentient creatures, and there is good reason to take Hamlet's opinion of himself as being applicable to each and every one of us:

> I am myself indifferent honest; but yet I could accuse me of such things that it were better my mother had not borne me: . . . What should such fellows as I do crawling between heaven and earth! We are arrant knaves all.
>
> (*Hamlet*, Act III, scene 1)

The case against the human race is fully appreciated by those who support the Green Party. Greens are concerned with saving all species, including mankind, but they are well aware that it is human activities that threaten the planet and all its inhabitants. In their book *A Green Manifesto*,

Irvine and Ponton describe human activity as a malignant cancer that can lead not only to self-destruction but to the destruction of other forms of life. We have already destroyed some 40 per cent of the Earth's productivity, thereby damaging its life-support systems, so that, as well as robbing our descendants, we rob other species of their means of life. They write:

> All other plants and animals work together in self-maintaining eco-systems, fitting into the biosphere as a whole. First with agriculture, now with the industrialization of one part of the Earth after the other, humanity is breaking apart the web of life.
>
> (Irvine and Ponton 1988: 40)

Although the developed countries have caused the most harm and many of our current attempts to reduce or to reverse ecological damage are at the expense of the poorer nations, there are regimes in the Third World that have been active accomplices in the maldevelopment of their lands. In addition they exacerbate their problems by virtually unrestricted population growth.

> There is a great deal to value and to learn from the cultures of the Third World. Yet we in the rich West do harm if we remain silent about the forces within the Third World that encourage environmental destruction, poverty and population growth.
>
> (Irvine and Ponton 1988: 126)

But as well as preparing the way to destroying ourselves, we have exterminated, and are currently exterminating, other species that have an equal right to live on the Earth. Of course species became extinct in the past, long before there were human beings, and new species took their place, but we are actively driving species to extinction.

Humanity stands indicted and if morality were primarily a matter of logic there could be no rejection of the conclusion that we have a moral obligation to exterminate our species. I am of course aware that such a view is based on a secular morality and that those who believe in a personal God will have arguments to bring against such a conclusion. I do not propose to discuss the implications of religion here, because the moralists to whose views I object do not themselves appeal to religion to support their arguments. In general it is fair to say that they were mildly hostile to religion and certainly they wish to establish their conclusions without invoking a deity. Thus Singer says:

> In discussing the doctrine of the sanctity of human life I shall not take the term 'sanctity' in a specifically religious sense. The doctrine may well have religious origin . . . but it is now part of a broadly secular ethic, and it is as part of this secular ethic that it is most influential today.
>
> (Singer 1979: 73)

My own argument is set out below:

> It is wrong to cause pain and suffering to any sentient creature and we have an absolute moral obligation to try to reduce it.

> The human race has always caused far more pain and suffering to sentient creatures than any other species and is so constituted that this is inevitable.

> There are far more non-human sentient creatures than human beings.

> Therefore we ought to encourage and advocate mass suicide.

I suggest that most poeple who are told that moral logic shows that it is a moral duty to urge the annihilation of our species will take the view either that logic is faulty, or that the facts have been misunderstood (so that at least one premiss is false), or that the problem of what is ideal moral behaviour cannot be solved solely by appeal to logical argument. I do not think the logic is faulty; what of the truth of the premisses? The first premiss is certainly true and is indeed simply a statement of an accepted moral position. The third premiss cannot seriously be disputed. Only the second premiss is debatable and mankind may not be quite such a disaster for the planet as I fear; nevertheless I do not think it would be easy to acquit our species of the charge of doing far more harm than good and I also think that there is a good reason to believe that this is inevitable. Even if we modify the second premiss and suggest that we might come to cause a little less misery (to ourselves as well as to animals) it will still support the conclusion. However, if we are prepared to accept a limited speciesism we can decide to modify the first premiss and to replace it by something like:

> It is wrong wilfully to cause unnecessary pain and suffering to any sentient creature and we have an absolute moral obligation to keep any necessary pain and suffering to a minimum.

I am prepared to accept such a modification because I do not regard a restricted speciesism as immoral; I rate the interests of human beings higher than the happiness of other sentient creatures.

Those who do accept some speciesism have sometimes been compared to racists and have been condemned along with racists. But the analogy is false. Speciesism has been distinguished from racism on the ground that there are objective differences in sentience between *Homo sapiens* and other species. Critics of this distinction say first that in the case of apes and certain other mammals the differences are barely greater than the differences between humans, and second that certain special classes of humans (infants, the senile and the mentally subnormal) are markedly less sentient than many animals. Hence, if degree of sentience is the criterion, these groups should have less consideration than many animal species. I think that these two criticisms can be overcome but perhaps a more significant distinction between speciesism and racism can be made. Racism is a form of xenophobia and is based on hatred and fear of strangers; it is motivated primarily by active malevolence, a wish to do down or keep under the alien race (or races). Admittedly this attitude entails relative elevation of the favoured race (or races) but this is incidental. By contrast speciesism is not motivated by hatred and fear of other species; the preference given to one's own species does not involve hating other species. On the contrary, one can favour *Homo sapiens* and still be fond of animals; and one can have more consideration for some animal species than for others.

The argument can be developed to show that it is not necessarily immoral to favour the happiness of those particular human beings whom we know and love. It is not unethical to be more concerned for their well-being than for the well-being of strangers. This is not to say that non-human sentient creatures and human strangers are not objects of any moral concern but to assert that they are and cannot be of such great moral concern as those who are close to us. This applies even to negative moral obligations. We are more concerned that those we know and love should not be harmed and should not suffer injustice than we are concerned for strangers even though we can acknowledge that our negative obligations to strangers are as strict. If the moralists claim that such a position is illogical, then so much the worse for moral logic.

Last we must accept that pleasure is an important part of all our lives; it helps to make life worth living. Therefore we all need pleasures and in this twentieth century many relatively simple pleasures require money. For example, most people in Britain acknowledge that the poor in our own society need not only food, warmth and shelter but also TV sets and the means to buy such comforts as a drink at the pub, newspapers and (possibly regrettably) tobacco; these are seen as necessities not luxuries. We must appreciate that it is not necessarily immoral to enjoy ourselves when others are suffering, even when we might help to alleviate that suffering by forgoing our pleasures. I have already suggested that readers of this book will be living in much greater comfort than that needed to sustain life and health. In addition most of them will indulge in 'treats': they may visit art galleries, possibly buying a picture or print; they may have drinks in a pub or bar; they may have meals in restaurants. Practically all of us conduct ourselves in this way though we are well aware that sentient creatures everywhere are in distress. Yet we are not, in general, perpetually oppressed by this fact and this is not, I suggest, because we are grossly selfish and immoral and would drink champagne whilst watching an *auto-da-fé* but because each human life (indeed the life of each and every creature) is lived in a world which is a world of individuals who can only relate emotionally to a small number of other individuals.

If our lives are to be anything more than a means of producing the next generation this has to be the case; quality of life is as important as life itself and is of moral significance. Indeed, in their discussion of euthanasia Singer and Rachels consider the quality of life to be more important than life itself. Most of us will agree: we think it important not only to have enough to eat, to be reasonably warm and to be in good health but also to have opportunities to develop social, physical and intellectual interests. As I said earlier, we might well be happier leading simpler lives and with fewer consumer luxuries; we should certainly be less of a menace to the planet and to the other species with which we share it; but even then, we should still require more than basic subsistence to make life worth living. Just as we are capable of joy even though we know that strangers suffer, so those strangers will be capable of joy when we suffer. Of course this is not because we suffer but because it is impossible to have active concern for all creatures all the time. Possibly this is not in accord with moral logic but it is in accord with the human feeling in which morality is rooted.

Such a view is not accepted by Singer for he maintains that there are no strangers and that we are, and in effect always have been, interdependent. Therefore we all have a right to expect help and we all have a duty to render help:

> If we consider people living together in a community, it is less easy to assume that rights must be restricted to rights against interference. We might, instead, adopt the view that taking the right to life seriously is incompatible with standing by and watching people die when one could easily save them.
>
> (Singer 1979: 166)

Now this account conveys a picture of a group of people living together who will be aware of each other as individuals and who will know of distress through direct personal appeal. I suggest that the mutual obligations in such a community are not to be conflated with those in the global village. As members of that village we of course have negative obligations. Apart from our duty not to be unjust we have a duty to refrain from killing, rape, torture and theft and a duty to refrain from reckless behaviour (such as dangerous or drunken driving). Our duties also include refraining from polluting the environment by contaminating air and water, and from inconsiderate behaviour such as parking in prohibited areas. Negative obligations cover a good deal of interpersonal behaviour and help to make us social beings.

For reasons already given, positive obligations must be restricted and must be specific. Positive moral obligations are cone-shaped rather than spherical, the base of the cone, entailing the broadest field of obligation, being sited among those who are emotionally close to us. We have positive obligations to all those whom we acknowledge as our dependants and such obligations may far transcend a duty to preserve their lives. We also have various positive obligations to friends and neighbours, and we may have explicit obligations (based on promises and/or legal contracts) to relative strangers. In addition we may have positive obligations to household pets and working animals in our care. But the degree of positive obligation is variable and is not absolutely categorical. One would risk one's life to save one's child (though, *vide infra*, such action is unlikely to be regarded as a matter of *duty*) but one might not feel obliged to risk one's life to save a stranger. Singer refers to the ease of giving help, and there is no doubt that if it is relatively easy to relieve another there is a stronger sense of obligation, but I suggest that one's relationship to the sufferer is of even greater importance. My main point, however, is that very often, and perhaps much more often than not, acceptance of a positive obligation is not just a result of assessing the moral logic of the situation. It is the result of an emotional response and is often spontaneous. We cannot therefore conclude that if we wish to help one particular person in a particular situation then it is morally obligatory to help all those afflicted in similar situations. Positive obligations are not general obligations. As Williams says:

> The point of negative obligations does lie in their being general; they provide a settled and permanent pattern of deliberative priorities. In the positive kind of case, however, the underlying disposition is a general concern, which is not always expressed in deliberative priority.
>
> (Williams 1985: 186)

I said that to rescue one's own child would not generally be regarded as a matter of duty (though it is very often a duty) for much of our behaviour to those in our care is motivated by love rather than by duty and there can be no clear-cut distinction between positive obligation and benevolence. The Good Samaritan can be commended for a neighbourly sense of duty and also for Christian charity. But if we are not neighbours and if (unlike the Good Samaritan) we are not directly confronted by suffering then what we may do to help strangers in distress is a matter of charity (in Saint Paul's sense). Any individual, and clearly Singer and Rachels are such individuals, may consider that they have a positive moral obligation to help but I surmise that everyone, including Singer and Rachels, will set some limit, above maintaining subsistence, on what they give.

However, although we have no general positive obligation of benevolence we do have a general negative obligation to refrain from injustice and therefore we have seriously to consider whether, as a matter of justice, we have moral obligations to the poor in distant countries. The question has been discussed by Onora O'Neill in her book *Faces of*

Hunger and it is from this book that I take my argument. O'Neill considers the notions of the rights of the needy, in terms of human rights, and here she disagrees with Singer's view that there are general obligations to them on account of such rights. She points out that

> Holders of rights can press their claims only when the obligations to meet those claims have been allocated to specified bearers of obligations. . . . Since the discourse of rights assumes that obligations are owed to specified others, *unallocated* right action, which is owed to unspecified others, tends to drop out of sight.

And she continues that 'When "rights" are promulgated without allocation to obligation-bearers they amount to empty "manifesto" rights, whose fulfilment cannot be claimed from others.' Of course Singer would claim that we in the affluent countries do, each and every one of us, carry an obligation but we have already given reasons for denying this. O'Neill makes the same point: 'Nobody can feed all the hungry, so the obligation to feed the hungry cannot be a universal obligation, and most of those who are hungry have no special relationship in virtue of which others should feed them.' She argues that if we focus on rights then generous impulses are likely to seem less important and in addition others' needs will tend not to be thought of as reflecting an injustice unless special rights can be shown (O'Neill 1986: 100–2).

It is the injustice *per se* which supports moral arguments for action. However, since human beings are not perfectly rational and can make mistakes they are vulnerable to deception and coercion – intended and unintended – and the general obligation to resist injustice must, O'Neill says, be supplemented by particular obligations (imperfect duties). These particular obligations will subsume benevolence and O'Neill asserts that any agent who is consistently non-benevolent must violate such obligations. Thus, though she regards justice as fundamental she sees beneficence as an integral part of justice. She says:

> Justice is the more fundamental obligation because it concerns the framework of institutions and practices. . . . However, because it is a matter of structure, justice cannot provide everything that human beings in specific circumstances need to be able to act. Even where material justice is in principle secured it may be lacking in practice. Amid unjust institutions there is no limit to the needs to which the imperfect obligations of beneficence and development of human potentiality may have to turn.
>
> (O'Neill 1986: 161)

O'Neill stresses that a failure of beneficence to bring about a just distribution, that is, to do the task of justice, is no failure of beneficence itself: 'Voluntary and charitable organizations do not fail in beneficence if they cannot meet all material needs, so leaving some still in desperate poverty' (O'Neill 1986: 162).

She suggests that we do have a moral duty to seek and promote justice through political action. She advocates campaigns to persuade those who govern to approach the issue in new ways and to refuse to construe the world's problems in terms of an approved grid of categories.

In this context it is helpful to consider the activities of those social reformers in England at the beginning of the twentieth century who were concerned with the desperate condition of the urban and rural poor. The Fabian Society was established before the Labour Party, and Beatrice and Sidney Webb were prepared to work with Conservatives and Liberals to further their objectives. Their activities were crucially important in establishing the principles of unemployment and sickness benefit and of old-age pensions. Yet most members of the society were comfortably placed professional people and many, such as the Webbs, Bernard Shaw and H. G. Wells, lived in real affluence.[2] They saw nothing immoral in this and there is no doubt that they achieved more in the long term for poor working people than did those who sent money (even considerable amounts) to religious and secular charitable funds. In so far as morality consists in trying to alleviate conditions political activity is far more effective than self-denial. Of course there is a place, a major place, for charity but we need to be aware that by temporarily mitigating hardship, charity may impede attempts at fundamental reform; almost certainly it will not produce long-term solutions for the problem of poverty. It is not irrelevant to point out that many political and social reactionaries have encouraged giving to charity, at least to the 'deserving poor'.[3]

What moves us to benevolence moves us by awakening our feelings of love and compassion for fellow creatures not by stimulating our powers of logical analysis. This is apparent if we consider the case of a person who filfuls all negative obligations and all specific positive obligations but never shows any spontaneous benevolence. It seems to indicate a certain coldness even though it can be argued that such a person might be saintly in that he or she regarded as a duty what others took as an option. But in fact a complete lack of spontaneous benevolence is very rare, whether it be subsumed into the saintly sacrifice (in the name of moral obligation) of all possessions and pleasures or whether it be totally rejected so that only strictly limited personal obligations are acknowledged. Most of us take a path between these two extremes and what we do for strangers depends on the nature of the appeals and our own feelings: some may give to Oxfam, some to the blind, some to political refugees, some to the local hospital and so on. A characteristic of benevolence is that it is based on a wish to give a certain kind of help to certain people or groups of people in certain situations. Any act of benevolence whether near at home or far away, whether towards human beings or other animals, whether to help save life or to improve the quality of life, merits moral approbation and encouragement.

To summarize: first, moral logic cannot establish a case for all-but-total sacrifice for those in need and there are no grounds at all for maintaining that the starving in far-away countries have prior claims over all other forms of relief. This is not to say that those who wish to make great sacrifices do not merit praise but to insist that there is no moral obligation to act in this way. Second, those who advocate self-abnegation as a moral imperative are ignoring important aspects of the human condition that make their recommendation impractical and indeed may distract attention from other actions that are much more likely to be successful. Third, it must be stressed that pleasure is a valuable part of each human life and human beings not only do live but also need to live without perpetual worry about suffering. Indeed I suggest that those who are overcome with a sense of guilt if they indulge themselves and who are filled with moral indignation when contemplating the pleasures of others are apt to lose their love of all their fellow creatures. True benevolence arises when it becomes part of one's pleasure to make others happy; there is Pauline charity when, as Mill says, the general happiness is part of each individual's happiness.

The world would indeed be a better place if we all showed more charity but though benevolence may mitigate it will not cure social evils. We must assess people as they are and seek cures; this is a moral as well as a pragmatic principle. Moreover exhortations that imply that benevolence is obligatory may do harm because, if there is no distinction made between benevolence and duty, we arrive at a wrong conception of the nature of moral obligation. This could not only discourage benevolence itself but also lessen the force of our moral obligation to reduce the injustice in the present distribution of resources. In one sense poverty will always be with us, for it is a relative condition, but we have the means to abolish destitution. We need the will.

NOTES

1 I am indebted to Professor David Walker of Exeter University for advice and guidance on this point.
2 See Holroyd (1989: chapters III, V).
3 See Doolittle's remarks in Bernard Shaw's *Pygmalion*, Act III.

3 Towards a more humane view of the beasts?

Mary Midgley

PUZZLING PREDICAMENT

Are we more or less humane towards non-human animals than our parents were? This may seem a rather surprising question. Undoubtedly, during the last few decades the more articulate among us have talked and thought much more than we used to about our responsibilities towards them, and have quite often acted on our new thoughts. What is unlucky is that, almost without our knowledge, at the same time our society's actual practical exploitation of them has been, and still is, steadily increasing. Much the biggest factor here has been industrialized food and fur production, which now processes huge numbers of quite advanced social animals away from public sight, treating them with total disregard for their natures and feelings, exactly as if they were lifeless objects. Life for animals on the traditional farm was of course often hard, but it did necessarily involve some degree of concession to their natural wishes – a concession which modern technology has now eliminated. I cannot cite details here, but in case anyone thinks this problem has already been contained, it is perhaps just worth mentioning the effects now beginning to flow from the recent US law allowing the patenting of newly bred animal life forms. Commercial research is now in full flood. Grossly large pigs have already been produced, and have proved to be chronic invalids suffering from arthritis and crossed eyes. Conditions such as these would not at all necessarily affect the meat value of animals, so they would not prevent their being produced. To quote – not from any animal welfare source, but from the *Meat Trades Journal* (London) for 30 April 1987 – we should expect

> in twenty-five years' time, truly bizarre farmyard scenes – cows with hugely distended rumps for meat; vastly extended pigs for prime back bacon; featherless chickens. With their anatomies so grossly distorted, the animals would function through life-support systems. In fifty years' time, 'tumour-farms'. A tumour is a swelling caused by the abnormal growth of perfectly natural tissues. *This would be a very efficient form*

of meat production, requiring little space and husbandry [emphasis mine].

In scientific experimentation the numbers are of course much smaller and some controls on their treatment do exist. But the situation is in general similar and in certain ways it may even be considered worse. For one thing, primates are still quite widely involved. For another, the nobility of science itself – its deservedly high moral standing as an activity that constitutes one of the main glories of our civilization – makes it all the more disturbing that it should use practices which cause less surprise in industry and commerce.

In pointing all this out, I am not taking up some peculiar, unrealistic, cranky personal standpoint. I am drawing attention to a discrepancy in the sensibilities of our age. The steady growth of callous exploitation is occurring at a time when our response both to individual animals and to nature as a whole is becoming ever more active and sensitive. There is accordingly now a much greater gap between the way in which most of us will let a particular animal be treated if we can see it in front of us and the way in which we let masses of animals be treated out of our sight than has arisen in any previous state of culture. Discrepancies like this between thought and fact are very significant in the history of morals. They produce deep uneasiness and real difficulty in considering practical questions.

An obvious case is the disturbance of feeling which arises in sensitive and sophisticated societies which have traditionally used slavery. I think that at present, as has happened in those cases, many people are beginning to bring their uneasiness about this particular anomaly to the surface. In particular, veterinarians are among those who are often struck by it. There is plainly something strange about the contrast between the anxious and responsible care that we often show towards our pet animals and our attitude to other very similar ones – for instance, pigs, which are creatures about equally sociable, active and intelligent with dogs and cats. In general, our current morality is essentially humane – that is, it takes suffering, and the infliction of suffering, very seriously. It does not simply dismiss pain and misery, as earlier moralities sometimes have, as mere trifles, or perhaps salutary trials, necessary and even valuable elements in earthly life. So, does this highly intellectual and morally conscious society have a quite satisfactory explanation for the difference between its ways of treating various kinds of animal?

THE DETRIBALIZATION OF CONTROVERSY

Simply to spot an anomaly like this, and to be disturbed by it, is in itself a useful experience. All moral advances come about through such painful recognitions of inconsistencies, producing in time the gradual, often reluctant admission that things will somehow have to be changed. Thus over slavery, questions such as 'why is a slave child to be treated so differently from an ordinary child?' gradually seeped through people's defences, and finally led to action. When this starts to happen, a welcome regrouping of social forces becomes possible, and I think this is already beginning to happen about animals. No longer is there a quite simple tribal confrontation between embattled defenders of all existing practice and attackers of it, who are automatically labelled as cranks. Instead, there begins a much more interesting and constructive debate about what is to be done next. Nearly everybody now agrees that some things are wrong; the question is where to start altering them and how far to go. But we do not need to wait till we have settled the question 'how far to go' before we can start. As with other reforms, that is a matter that can be settled later.

My impression is that, about our treatment of animals, public opinion is moving quite fast in this more constructive direction. I am particularly cheered to notice this happening among scientists, more especially among the younger ones. Instead of automatically closing ranks against any criticism of objectionable experimental methods, an increasing number of scientifically trained people are now themselves uneasy about such things, and are prepared to take the initiative in reforming them. Instead of dismissing every criticism as the invention of anti-scientific fanatics, many of them have themselves noticed that inhumane experimental techniques are often unnecessary and sometimes scientifically faulty. Moreover, they can see that, for the credit of science itself, it is important to avoid all well-grounded imputations of the use of brutal methods. Since the general public also now contains a much wider range of well-informed critics than it used to, this makes possible a less confrontational and more co-operative approach to reform. Instead of dividing on the stark alternative of 'no experiments' or 'all existing experiments' we can now more easily work to make existing experiments more humane, to cut down their numbers and to rethink the distorted intellectual background which has sometimes made excessive numbers of them seem necessary. Over zoos again, we can avoid the stark conflict – 'all existing zoos' or 'no zoos' – by pointing out the difference between good and bad methods of zoo-keeping and working to make bad zoos better, or indeed when necessary to close them down.

In this sort of way, by avoiding simple tribal conflict, it often becomes possible to spot ingenious alternatives that will greatly alleviate existing evils. For instance, there is Jane Goodall's current Chimpanzoo Project in the USA. This scheme brings together chimps from a number of zoos out of their concrete cells into bigger colonies with more natural conditions – incidentally providing material for aspiring ethologists, but primarily getting the chimps a respite, to serve the rest of their undeserved life sentence in what is effectively a comfortable open prison. Since it is not possible to return such zoo-addicted animals successfully to the wild, it is hard to see what better fate could be found for them. Similarly, about meat and other animal products, there are many ways in which animals can be less miserably kept, even if they are still going to be killed in the end. Since there is no way to make everybody instantly turn vegetarian, and since many people are now willing to pay a little more for more humanely produced meat or eggs, this is a promising, immediate way of reducing the mass of needless misery.

THE MEANING OF COMPROMISE

For some reformers, such methods are too slow. But the question for us surely is: if you were a non-human animal, caught up already in our vast and terrifying social machine, would you want the people campaigning on your behalf to promote schemes like these? Or would you prefer that they should refuse all such half-measures and wait instead for the moment when the whole disgusting business can be swept away by a single cleansing explosion? From the animals' angle, the explosive option does not look persuasive. Those who demand such options see them as moral imperatives, principles that rule out compromise and are valid independent of all pragmatic judgements about the expected effects. But this line seems really to draw much of its force from a belief about such effects – the belief that (as Marx thought) slower, more piecemeal methods of reform cannot work. In fact, the history of such reforms as have been successful shows that they can, and indeed that they have done so much more often than attempts at sudden, total revolution. In particular, when what has to be reformed is a subtle and sophisticated institution, such as the scientific practice that surrounds experimentation, there is no way of producing useful change without the co-operation of those who understand its current working. That is why it is so essential both that scientists should take part in that work, and that outside campaigners should be willing to work with them when they do.

In some ways, then, the extremist, no-compromise approach to reform is mistaken. But it has an importance and a value that should not be forgotten. It makes an essential point about mood and attitude. It tells us that much that goes on is horrible. This is true, and accordingly, as with other reforms, when we accept a compromise, we ought to know that it is only a compromise and to continue to look beyond it. Over human affairs, this large demand is uncontroversial, because it is taken for granted that human beings ought to get a much better life than most of them do, so that it is always appropriate to aim at a high and distant ideal, beyond what one expects to achieve immediately. But about other creatures, not everybody finds this obvious. There are those who say that we ought not really to be concerned about anything except human beings. And of late an attack has also been developed from the other side, from those who say that concern for animals is too narrow rather than too wide, because what we ought to be concerned about is the whole biosphere.

WHO ARE OUR NEIGHBOURS?

The question 'what kinds of being do we owe duty to?' has thus been thrown open very usefully for new consideration. In a way, the two attacks just mentioned balance each other neatly. But they cannot just be left to cancel each other out. Both of them have a point, and their combined effect is to call for a radical rethinking of our whole relation to the planet we live on. This rethinking is now under way; it obviously needs a lot of attention from all of us if it is to take effect in time to save the world from serious trouble. What may be called 'exclusive humanism' – the belief that only human beings matter – has been deeply rooted in our cultural traditions and linked with much that was of great value in them. From the start, it was present in Christianity, which went to some trouble to narrow down older Jewish ideas of community with the rest of nature. Protestantism carried this narrowing process still further, clearing the stage for a drama involving only man and God. But it was taken further still by the Enlightenment – by a whole scheme of seventeenth- and eighteenth-century thinking centring morality on respect for Reason, conceived as a power possessed by man alone. This kind of Reason, as was repeatedly made clear, was not at all necessarily shared even with woman, and certainly not with any beast. Even God now withdrew from the stage, and western man was left there alone, to admire his own intellect in splendid isolation.

This tremendous concentration of interest on human intellecual achievements is still so central to our thinking that we have difficulty in criticizing it at all, let alone in doing so fairly. But in the last half-century, its shortcomings have grown increasingly glaring. Of course what is good about humanity must be celebrated and cherished. But the attempt to do this by sacrificing everything else – by cutting humanity off from its context, by rejecting everything non-human and non-rational as valueless – no longer looks today even like a rational project, let alone a realistic one. This change in our attitudes is still far from complete. But it is surely taking place right across the range of ideas that we live by.

On the physical side, such things as the destruction of forests and the pollution of seas are beginning to force themselves on the attention of even the most protected among us. No one who attends at all seriously to what goes on in the world can any longer treat human technology as omnipotent, though some economists still make a brave attempt to live in the past about this. And on the psychological side, the crippling narrowness of regarding a human being as essentially just an intellect has become terribly plain. These large changes can only be glanced at hastily here, but perhaps what follows for our present topic is clear. Concern for the suffering of animals, which used to be treated in our society as an outlying, rather eccentric interest, disconnected from the official morality of the tribe, now appears against a much wider background where it does not look eccentric at all. It is no longer obscured by the notion that humans are the only earthly things that have value, and that their value resides only in one special selected element – first the Christian soul, later the intellect. These are themselves highly eccentric moral ideas – indeed, fanatical ones which, once they cease to be taken for granted, would be very hard to defend.

In present-day morality, these mean and arbitrary notions have largely been replaced by two which are far more favourable to animals – namely, utilitarianism and the sense of community with the rest of nature. Utilitarianism is the view that the business of morality is essentially the promotion of happiness and the relief of suffering. Because animals can suffer as well as people, utilitarians have, right from their rise in the eighteenth century, included them unhesitatingly in the province of morality. As for the sense of unity with the rest of nature, it seems to be something which is probably necessary for our psychic health. The failure of our recent tradition to provide for it is no doubt one reason for the present popularity of various eastern religions, which commonly do give it its due. Yet this sense of unity is no stranger to our own tradition, as Artistotle and Hegel, Blake and Wordsworth, Darwin and Walt Whitman can testify. On the contrary, it is something which has recently been damagingly squeezed out of it by the excesses of exclusive humanism. Both Judaism and Greek thought fully recognized its importance, and it was actively present in the Catholic tradition, before Protestantism and some other factors drove it underground. It has also, of course, very strong scientific roots, being entirely congenial to the theory of evolution, and incidentally much stressed by Darwin himself.

SCIENTISTS AND THEIR KINDRED

The position of scientists on this matter is extremely interesting. Modern biology gives them plenty of reason to emphasize the continuity between humans and other animals, and to rebuke exclusive humanistic arrogance. Since Darwin's day, much detailed neurological work has confirmed how close this kinship and continuity really are. Indeed, if scientists did not believe this likeness to be very close, their reasons for a great deal of experimentation, notably on primates, would never have arisen in the first place. Perception of all this tends to favour consideration of the interests of experimental animals. On the other hand, in the past, the eighteenth-century exaltation of the intellect as the highest human faculty played a prominent part in encouraging the growth of modern science, and that exaltation still provides it with its simplest justification for existing, and for demanding funds.

Over the treatment of experimental animals, then, these two sets of considerations clash. It is important to reflect on just how far each of them is a necessary part of the scientific attitude. All of us, but especially scientists, need to think quite hard about the meaning of both these ideas, and – if we decide still to espouse both – to work on finding ways to reconcile the valuable elements in both of them. Darwin, who hated vivisection, and Claude Bernard, who defiantly gloried in it, typified the extreme positions that were taken up about this in the nineteenth century. The resulting feud was disastrous, but the bridge that should resolve their differences has not yet been fully built.[1]

Yet today it is becoming clear that we shall somehow have to build it. Bernard's ostentatious dismissal of the whole issue is no longer a position that makes much moral sense to us, more especially since the development of anaesthetics has removed the main practical consideration that made him adopt it. Conscientious western people today simply do not, in most of their lives, take the view that it does not matter at all what cruelty people inflict on animals, even though that view would be taken in some other cultures, and was defended in our own by philosophers as recent as Spinoza (*Ethics*, Part IV, proposition 37, note 2). And this change is not just a matter of fashion. It is part of a general shift in values, a conceptual revolution in which (as mentioned earlier) humane conduct and a sensitive attitude to suffering have become central elements in our moral world. Accordingly, people who justify inflicting this kind of suffering today are likely not to dismiss all objections out of hand in the way that Bernard and Spinoza found so convincing, but to give a much less sweeping justification in terms of priorities. It is admitted that animals have some claim on our consideration, but their claims are held to be in general outweighed by human ones.

HOLDING THE SCALES

This is a more flexible approach, so it needs to be worked out in greater detail. We ask at once: is this always so? Does every human claim, just because it is human, take precedence over every claim of another species, or is there some sliding scale? For instance, Eskimos eat seals just as the seals eat fish, and kill polar bears just as the bears might kill them – because that is their only way to stay alive. And in that equal struggle we may well back our own kin. But do we have to extend that backing to cover warm and well-heeled humans who simply happen to fancy wearing a sealskin coat, or having a stuffed bear in their hall? It is not in the least obvious why we should be expected to do this. We do not, after all, have to excuse every bad act committed by even our quite near human relations just on the grounds that they are our relations.

Defenders of existing customs do, however, often seem still to rely on a strong principle of human priority, which would in all cases make us send other animals to the back of the queue. Thus, people campaigning on behalf of animals are accused of heartlessly wasting resources which ought to have been used for humans. Indeed, concern itself is sometimes treated as a scarce resource, a fund on which only humans are entitled to draw.

It is interesting to notice that almost any useful action in the world can be resisted in this way, on the ground that there is something else still more useful which ought to have been preferred to it. Thus, if we propose to help a certain group of human adults, it may be objected that their children's need is greater, so why do we not start with them? If we do this, we can be blamed for not starting with the old, or the blind, or the handicapped children, or perhaps the specially gifted children who will be most able to help the others . . . and so forth.

There is a slightly mad mistake here which deserves attention. It lies partly in ignoring the fact, already mentioned, that we have to start somewhere. But what makes things much worse is the assumption that each of us only has only a very limited amount of help to give – say, a pint – and that this pint may not be poured out till we find a case that has been proved to have absolutely top priority. Since we can never be sure that we have found this special worst case, the effect is to stop us doing anything. Now even if our resources were indeed limited to this single pint, it would plainly be a much better idea for each of us to pour it out on the first urgent case that we saw. But actually, of course, our resources are not limited in this way. Every society that is a going concern at all has great reserves within it for those who need help. Most societies, indeed, habitually waste a great proportion of their resources and use some of them on things that are downright noxious. As for concern, like other feelings it is something that grows and develops by being deployed, like our muscles, not a sort of small oil well that will run out shortly if it is used at all. In fact, history shows that most of the major campaigners on behalf of animals have been people who also campaigned on behalf of unfortunate people – people such as Wilberforce, Bentham, Mill, Shaftesbury and Bernard Shaw.

It should be clear, too, that the idea of necessary competition between all those in need is mistaken. Helping one group often helps another as well – as it does, for instance, in the case of the adults and children just mentioned, who can certainly not be helped apart. Another good example is vegetarianism, which is not, as people often suppose, a sacrifice of human interests to animal ones. Most of the food that is now fed to food-animals to produce meat could be eaten by humans directly, and if it were so used it would nourish enormously more of them than it does by following its present indirect route to feed only a few. Human health, too, is not half as well served by meat eating as used to be supposed.

All this is typical of the ways in which the interests of different groups can converge quite as often as they can be in competition.

Of course we often do have to choose between needs that actually do conflict, and to use various principles to help us decide which of them is, at the time, the more urgent. But the weakening of exclusive humanism removes the very simple pair of scales on which these questions used to be weighed. It is no longer just obvious that every human interest, however weak, outweighs every other interest, however strong. We are left now, not only with a rather different set of principles, but also with grave doubts about the competitive model itself. There is reason to suspect that we have been far too ready, in the past, to assume that conflicts of interests could not be resolved by reconciliation. The new ecological approach is gradually hammering into our reluctant urban heads the truth that interests are not always in competition. Adam Smith did not get it right. The interests of animals, or indeed of people, do not in general compete with those of forests or of rivers, but are interdependent with them. Animals need habitat and habitat needs animals. Even that most awkward and intrusive life form, *Homo sapiens* himself, is part of an ecosystem and cannot get by without it. The project of pursuing his interests on their own, in competition with the rest of creation, has been tried and has been proved self-defeating.

NOTES

Acknowledgement: Parts of this chapter have appeared in a pamphlet called *Conflicts and Inconsistencies over Animal Welfare* published by the Universities' Federation for Animal Welfare (UK) (Hume Memorial Lecture, November 1986), and I am grateful to the Federation for permission to reprint them.

1 For the views of Bernard and other protagonists in this feud, see Clark (1984) – editors' note.

4 The ethics of tourism

Robert F. Prosser

The jingle which envelops you as you drift through one of Disneyland's favourite attractions repeats insistently that 'It's a small world after all'. The intended message is of the 'oneness' of humankind. A more cynical interpretation is that it echoes a marketing manager's credo – 'OK, folks, it's a small world today, so let's get out there and enjoy it!' Technology in the form of high-speed, broad-bodied jet aircraft, computerized information systems and sophisticated media-based marketing, combined with increased affluence and time availability, have made leisure and tourism one of the world's fastest growing industrial sectors. From the 1960s until the mid-1980s, it was growing at 5 to 10 per cent a year in monetary terms, and despite a recent slow-down it is now the third largest industrial sector in the world (Figure 4.1). Worldwide, at least 10 million people work in tourism and many more millions live off tourism indirectly.

In industrial and post-industrial societies, travel has become a social norm. What has until recently been a phenomenon of rich western countries is now exploding across other societies and cultures. For instance, at the opening of the 1990s, some of the most rapidly expanding origins of tourists were the Asiatic countries of the Pacific rim. The Japanese are bringing their energy and spending power to the world of mass tourism. None the less, it is well to remember that over 60 per cent of all international travel still originates in North America and western Europe. Travel origins are today global, but equally, so are destinations, for over the past forty years or so the 'pleasure periphery' has rippled across the world: an aircraft carrying tourists has crashed in Antarctica; Mount Everest is polluted by tourist debris; Alaska's grizzly bears are coming to terms with the glint of camera lenses; peoples of Papua New Guinea adapt their crafts to the tourist market.

Until relatively recently, this expansion in holiday entitlement and holiday taking was greeted with all but universal enthusiasm. It has been used as a key indicator of improving living standards and quality of life for increasing millions of people. The 1960s 'saw the beginning of a universal and unrestrained tourist development euphoria' (Krippendorf 1987: 68), and grandiose claims were made; in 1961, Hunziker (1961: 2)

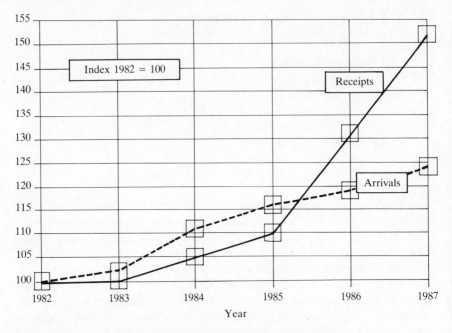

Figure 4.1 International tourist arrivals and receipts (world)
Source: World Tourism Organization.

wrote: 'Tourism has become the noblest instrument of this century for
achieving international understanding.' By the mid-1970s, different perceptions were beginning to surface, and 'the landscape eaters and the all
too obvious effects of tourism on the landscape had become an issue of
public debate' (Krippendorf 1987: 68). Turner and Ash, in their book
The Golden Hordes (1975), mounted a powerful critique of the mass
tourist and the tourism industry. By the end of the 1980s, a wide-ranging
reappraisal of tourism was taking place in both origin and destination
countries across the world; e.g. organizations such as Tourism Concern
were active in seeking alternative approaches, and terms such as 'soft
tourism' and 'eco-tourism' became enshrined in the jargon.

  This concern and disillusionment seem all so sad, as holidays are meant
to be the high point of our leisure lives, planned and anticipated keenly
throughout the preceding year; e.g. as we sit soporifically absorbing our
Christmas lunch, we are suddenly bombarded from our TV screens with
summer holiday adverts! If, by the end of January, we are not boring
our friends with the minutiae of our holiday planning, then we are
regarded as somewhat eccentric. Psychologists and sociologists make a
range of positive claims for the benefits of travel and holidays, related
to motivations, need satisfactions, education, social skills, etc., as this
list of section headings from a recent book illustrates:

Travel is recuperation and regeneration
Travel is compensation and social integration
Travel is escape
Travel is communication
Travel broadens the mind
Travel is freedom and self-determination
Travel is self-realisation
Travel is happiness

(Krippendorf 1987)

Through our holiday travel, we distance ourselves physically and psycho-
logically from our everyday lives. Our motives are essentially 'going away
from' rather than 'going towards', and hence our behaviour and motives
become markedly self-oriented. Such motivational biases become important when assessing tourism impacts upon the destination society and
environment.

When on holiday, our behaviour may change not simply because we
have time available but because we are removed from the norms and
strictures of our everyday lives – we are 'free'. We may give little thought
to whether such behaviour is socially or environmentally acceptable in
our chosen holiday destination, and indeed, if we are aware, we may
rationalize that we have paid and 'they' have taken our money, so we
are entitled to enjoy our holiday as we wish. This scenario encapsulates
one of the central dilemmas of tourism: is it, by definition, a selfish and
self-indulgent experience? If so, then is it realistic to expect significant
change to be energized from the 'demand' end of the system?

Theories of leisure and tourism emphasize concepts such as 'freedom',
'autonomy' and 'choice'. In popular parlance, it is all about 'doing my
thing'. But how 'free' are we? The formidable marketing skills of the
promoters of tourism are focused upon us, manipulating our minds,
directing our choices, channelling our 'freedom'. This scenario suggests
that any evaluation of the ethics of tourism should begin not from the
tourist, but from the tourism industry. In one of the most cogent critiques
of modern tourism, Krippendorf sums up the hard-sell, free-market
approach succinctly and bluntly: ' "Tourism is business, not charity",
"To hell with paradise!" is how promoters of tourism talk . . . why a
journey is undertaken is of no consequence to them – what matters is
that it is undertaken' (1987: 20). Yet there is little doubt that without
the hard bargaining, the sophisticated organization and the economies of
scale adopted by tourism developers and tour operators, many holiday
destinations would remain beyond the pockets of the majority of people.
In this way, the tourism industry enhances the range of choice and
potential experiences.

There are clearly two dimensions to the ethical debate about the
tourism industry: first, the methods it uses to manipulate and stimulate

demand; second, the strategies used to organize and make available the supply of holiday opportunities at the destination end. Selling holidays is all about selling images: 'Imagine yourself on a tropical island'; 'Ahead lie the snow-capped peaks of the Karakoram'; 'Meet new friends and dance the night away'; etc. Despite the strengthening of marketing standards legislation and improved levels of responsibility among the operators, there remain considerable differences between the image and the reality of tourist experiences. The photograph in the literature put out by Cook's, Thomson or Intasun shows a handsome young couple alone on a curving beach fringed by palm trees and an empty azure sea. What it may not show is the 500 room hotels backing the beach and the horde of guests occupying them. Nor does it show the urban slums of the nearby town. Indeed, why should it? For it is selling the 'Let's get away from it all' image central to the motivations of many tourists.

In the case of mass package tourism, the marketing message is so strong, and the scale of arrivals and impacts at the destinations so huge, that the components of the ethical debate can be easily identified. Yet the elitist and even arrogant attitudes and perceptions which underlie our burgeoning search for holiday experiences may be observed, albeit more subtly, within the sophisticated 'sales pitch' of top-of-the-range tour operators. The basic indicator of this syndrome is that we are wooed as 'travellers' not 'tourists'. We are offered exciting insights into fresh worlds which few others have experienced – but with all the familiar creature comforts which befit our superior status.

Analysis of even 'high-class' tourism marketing literature may be used to illustrate this conceptualization of travel as an amalgam of educational, spiritual and sybaritic experiences. Let us take the example of advertising for Society Expeditions, which appeared in January 1990 in *World* magazine, a coffee-table glossy environmental and travel monthly out of the same stable as *Geographical Magazine*. The experiences on offer are undoubtedly of high quality, and it may be argued that by their character and cost they involve small numbers and minimal impacts. By such criteria they may be said to represent 'responsible' tourism. Yet consider the perceptions and images implicit and explicit within key words: 'high-class' tourists become 'discoverers' on an 'expedition'; they visit 'lost', 'timeless' islands, 'undisturbed' and 'untouched' ecosystems, 'primitive' (*sic*) and 'mysterious' peoples. Are we not intruders, as we arrive in our air-conditioned environments with 'gracious and intimate' service? (Are the 'Europeans' the officers and the 'Filipinos' the 'dedicated' crew?) How can an island remain 'lost' and 'undisturbed' when we and our '139 other expeditioners' arrive regularly? What are the messages we receive and accept concerning stereotyping and superiority/inferiority, within terms such as 'primitive' and 'mysterious peoples'? Conversely, genuine efforts are made to enhance human and environmental understanding via the educational and interpretation elements of the package.

It is within such experiences that the true dilemma of tourism may perhaps be glimpsed. On the one hand lie the positive human characteristics of curiosity, intelligence, empathy, enjoyment, adventure, fulfilment, all part of laudable desires to broaden our horizons, enhance our quality of life, etc. On the other hand, can we be sure that the destinations we probe by our power of mobility, affluence and time have an equal power to say 'yes' or 'no'? It can be argued, of course, that at the scale of tourism adopted here, the quality of the experience for the tourist and the quality of life and environment at the destination can be conserved. But this is in itself an elitist argument: why should only the rich be able to experience 'timeless' islands? If the tourism industry is an effective force for egalitarianism, and the model of tourism dynamics of Table 4.1 is allowed to evolve, will it be long before the cruise ship carries not 139 'expeditioners' once a month, but 1,500 'packaged products' once a week? After all, not so long ago, east Africa and Nepal were 'lost' and 'timeless' – now they are part of the mainstream tourist circuit:

> In Africa many of the Masai charge tourists for the privilege of taking their photograph, and cheap, quickly produced replicas of traditional art are sold. . . . While the tourist may only pay a few cents or pennies for such items, to the trader this is a considerable income. . . . In Nepal the boom in trekkers visiting the region has had a particularly dramatic impact; many villagers have burned their precious wood supplies to heat the tea urns. More and more these people are becoming reliant upon the trekking parties for their income, even abandoning their traditional agriculture.
>
> (Mears 1990: 86–7)

Table 4.1 A typology of tourists and their impact

Type	Numbers	Impacts
Explorer	Very few	Accept local conditions
Off-beat adventurer	Small numbers	Revel in local conditions
Elite	Limited numbers	(a) Demand western amenities (b) Roughing it in comfort
Early mass	Steady flow	Look for western amenities
Mass charter package	Massive numbers	Expect western amenities

Equally powerful ethical questions surround the organization of the supply of holiday opportunities at the destination end of the system. At a conference held in the Commonwealth Institute in London in 1990, a director of American Express set out the various ways in which the travel industry is today closely scrutinizing developments and impacts. When pressed by questioning, however, he admitted that 'the bottom line' for a company is the profit motive. If tour operators are to give holiday makers value for money and to compete successfully with their rivals, then they must strike hard bargains with hotels, bus companies, etc.; e.g. the accommodation element of a hotel-based holiday may comprise between one-third and one-half of the total expenditure at the destination. In turn the hotel owners attempt to keep down the construction and hence loan repayment costs, and the running costs. It is this focus upon economic returns which has led to much of what has been called the 'architectural pollution' (Mathieson and Wall 1983) of many low-quality and aesthetically mediocre tourist developments, e.g. the Spanish 'costas'.

It is, of course, dangerous to generalize, for tourism is a complex phenomenon. There are, indeed, many types of tourism – and tourist. The travel industry thinks and plans in terms of market segments, using variables such as numbers, price, activity/experience, destination, age, socio-economic status, accommodation type, timing and length of stay. Within this multivariate structure, two main categories may be identified: elite tourism and mass tourism. Each category has a distinctive set of requirements and hence impacts upon a holiday destination. These impacts are usually classified as economic, social/cultural and environmental, and analysed in terms of benefits and costs. The model of Table 4.1 (p. 41) summarizes the relationship between tourism, the tourist and the impacts. This model may be applied to different destinations at a given moment in time, or perhaps more importantly, to the evolution of an individual destination over time. For example, the explorer or adventurer traveller discovers a Pacific island paradise and passes the word around; the travel industry becomes interested in this potential new market, and makes it available to the wealthy, mobile elite. Then, by the process of successive class intervention, and by improved accessibility, increasingly mass arrivals are organized. By this stage, the adventurers and elites have moved on, pushing the pleasure periphery farther out.

Clearly, the economic, social/cultural and environmental benefits and costs to the destination escalate as this dynamic model evolves. If we turn our attention to this destination end of the system it is not difficult to identify why a country, especially a nation urgently seeking economic growth and development, should encourage the tourism industry. It is a high-growth industrial sector, and demands resources which seem readily available in many countries – sun, sea, sand, exotic cultures, beautiful environments, wildlife, etc. Governments of host countries see tourism as a fruitful source of foreign currency, employment opportunities, investment potential, tax revenue, regional development, infrastructure improvement, enhanced prestige and image. From Spain to the Seychel-

les, from the Caribbean to the Cook Islands, there are numerous examples of nations seeking this bonanza.

The prizes, however, have often been less than glittering. The problems most commonly stem from the issue of who controls the development and running of the tourism. Because of their limited ability to raise
investment capital internally, the absence of a well-developed construction industry and a poorly trained labour force, many countries have
approached or have succumbed to the blandishments of multinational
companies and foreign investors. All too frequently, therefore, the local
people progressively lose control of the activity. This loss of autonomy
has proved one of the most negative effects of the coming of tourism,
especially in small and vulnerable societies such as island communities.
For instance, in the early 1980s Club Med built a tourist village on the
Turks and Caicos Islands, with the aid of a grant from the British
Overseas Development Agency. Local people were enthusiastic at the
outset; the 7,500 population saw rosy job prospects:

> Those who will be either directly or indirectly employed immediately
> are estimated to number about 200. Over a five year period such
> employment is expected to have spread to some 600 people. And if
> . . . further hotel development follows, as many jobs again might be
> created over the next ten years.
>
> (*Overseas Development*, Feburary 1981: 5)

By the late 1980s, the TCI government and people were complaining
that only the lower-paid and menial jobs were being made available to
locals. Where tourism is controlled by foreign companies, the 'leakage'
of tourism-generated income may be as high as 80 per cent – e.g. profits
to go the foreign companies; expatriate workers send money home; food,
gifts, etc. are imported rather than produced locally (Figure 4.2). Yet
there is no doubt that tourism is capable of yielding considerable benefits;
e.g. it is by far the largest source of Spain's foreign currency income.
The next question focuses upon the distribution of such income. Evidence
from countries of all sizes, especially in the developing world, suggests
that only a small elite reap most of the gains – the rich get richer!
Articulating an effective system for spreading this wealth directly or
indirectly through all strata of a society is proving one of the more
intractable dilemmas of development policy, both internally and internationally.

A number of variables influence the character and scale of the social
and cultural impacts of tourism upon host communities. The two most
powerful are the *numbers* of tourists in relation to the host population
and the *cultural distance* between the 'guests' and the 'hosts'. As the
model set out in Table 4.1 indicates, the relatively small numbers of
'explorers' and even 'elites' may experience a destination without profound impact. However, once the visitor numbers reach certain critical

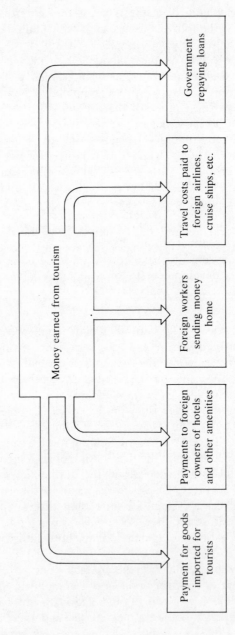

Figure 4.2 The leakage concept

mass – different for each destination – then impacts upon attitudes, perceptions and lifestyles accelerate. For example, many Mediterranean coastal villages such as Ayia Napa in southern Cyprus have long histories of small numbers of independent travellers passing through. These were welcomed and accommodated. Gradually, more families began to let rooms, café owners increased their trade of local food and drink, crafts-people were delighted to find that the travellers bought their products. At this scale, tourism was integrated within the economy and society. The coming of institutionalized mass tourism, whether organized by local Ayia Napans or by foreign companies, changed this relationship funda-mentally. Antagonisms developed between those who were benefiting and those who were not, and between generations. Younger people in particular have been influenced by the lifestyles paraded before them by the apparently affluent tourists. This 'demonstration effect' is one of the most commonly noted accoutrements of tourism, especially when there is a wide cultural distance between guests and hosts. Older generations find the styles of dress and sexually explicit public behaviour of the tourists offensive, and are angered when their younger relatives copy such fashions. Thus, in Ayia Napa, today one of the largest tourist destinations in Cyprus, tourism has come to dominate the physical struc-ture, the economy and the cultural patterning. The response of the local community has been to use some of the undoubtedly considerable income to build a new village away from their traditional home, in which they hope to be able to recreate their traditional way of life, yet continue to 'service' the tourist Ayia Napa.

The process by which tourism progressively takes over existing settle-ments is known as 'penetrative' tourism, and is common throughout the world, e.g. the transformation of the village of Kuta on Bali from a stop-off on the 'hippie trail' of the 1960s to a mass tourist venue of the 1990s. It arouses especially strong arguments concerning cultural ethics. On the one side there are those who see the commercialization of the artefacts, dances, temples, clothing, belief systems, etc. as immoral and destructive. Rising land prices force people out of their homes; the agricultural bases of life disintegrate; and social mores and cohesion weaken. On the other side, it is argued that tourism infuses new life into dying cultures, gives added pride to them by display and brings welcome income, hence encouraging young people to learn the traditional crafts etc. and to stay locally to earn their living. Examples are quoted to support both argu-ments: on Hawaii, the traditional dances, artefacts, clothing and host–guest relationship have been reduced to commercialized cabaret; on Bali, it is claimed that much stronger cultural identity and authenticity have been retained, and that the displays put on for tourists still have genuine cultural meaning. There is no doubt that fundamental changes do take place in the way that host populations respond to tourism. Doxey (1975) has produced a model of what he calls tourist irritation (Table 4.2),

Table 4.2 Index of tourist irritation

---

(1) *Euphoria*
* Enthusiasm for tourist development
* Mutual feeling of satisfaction
* Opportunities for local participation
* Flows of money and interesting contacts

(2) *Apathy*
* Industry expands
* Tourists taken for granted
* More interest in profit making
* Personal contact becomes more formal

(3) *Irritation*
* Industry nearing saturation point
* Expansion of facilities required
* Encroachment into local way of life

(4) *Antagonism*
* Irritations become more overt
* The tourist is seen as the harbinger of all that is bad
* Mutual politeness gives way to antagonism

(5) *Final level*
* Environment has changed irreversibly
* The resource base has changed and the type of tourist has also changed
* If the destination is large enough to cope with mass tourism it will continue to thrive

---

Source: Doxey, 1975.

summarizing the dynamics of perceptions. The universality of its application is, of course, open to debate!

There is, however, a second and quite distinct type of tourist destination – the custom-built resort. The package tourists are transported to their 'environmental bubble' where every aspect of their holiday experience is provided. On Bali, the massive Nusa Dua complex exemplifies this form of development, built intentionally away from existing settlements. The extreme example is the new Waikaloa Hyatt resort in Hawaii. This expensive, exclusive resort has been located on an empty expanse of barren lava flows, and has everything from golf courses to an internal computer-run boat transport system. All conceivable leisure experiences are brought on-site to the tourist: 'native' dancing, music and cookery displays, shops, a full range of water sports, game hunting, tactile experiences with dolphins, etc. This is 'enclave' or 'ghetto' tourism in its futuristic form – at a price. At the other end of the price spectrum, it can be argued that the hedonistic sun-sea-sand-sex experience of the Costa Brava or Majorca has similar characteristics, i.e. minimal contact with the host society, except in purely economic terms of jobs. A strong case can be made for seeing this ghetto and environmental approach as the only effective and realistic way of controlling the impacts of mass tourism. By isolating the facilities and the tourists from the mainstream of the host nation, the hosts may reap the benefits without enduring the worst of the costs.

Irrespective of the type of tourism, however, the industry has been criticized as inflicting a form of economic colonialism upon the host countries, many of which are recent politically independent nations. Wealthy and powerful foreign companies have replaced foreign governments while apparently and comparatively wealthy foreign travellers have replaced the administrators and soldiers. Especially where the control of the tourism industry is in foreign hands, tourism is seen as perpetuating the dependency cycle so prevalent in the colonial era. For instance, in Fiji, with a large tourism industry, much of which is controlled by international companies, the leakage of tourist money approaches two-thirds. In Tonga, on the other hand, the tourism industry is much smaller but is almost entirely in local hands, and the leakage is less than 10 per cent (Table 4.3).

Large-scale tourism inevitably has great environmental impact, both in the destruction of natural landscapes and in the development of built environments. As mentioned above, the worst excesses of 'architectural pollution' are today being avoided – e.g. building regulations enforcing the setting back of buildings from beach fronts; height restrictions; use of vernacular styles and materials. Yet any such developments take up space, and even where existing settlements are not removed, habitats, often of high quality, must disappear. In Australia, the north Queensland coast is at the beginning of a tourist explosion, energized by capital and tourists from around the Pacific rim. Apart from the attractions of the climate and beaches, there is the jewel of the Great Barrier Reef. As with any coral reef, this is a very fragile ecosystem, and it has been designated a Marine National Park. A critical management problem, however, is that the park does not extend to the mainland coastal strip. Thus, resort developments are mushrooming along the coast, most with marina or commercial boat facilities, whose key attraction is access to the Barrier Reef. Furthermore, the building of these resorts is destroying the previous mangrove forests of the coastal strip. Apart from their ecological value, mangroves also help to enrich the reef environment as many of the reef inhabitants spend part of their life cycle in these mangroves. In other instances, however, tourism can justifiably claim to possess significant conservation powers. It is unlikely that the great animal assemblages of the east African savannas would survive unless they were acknowledged and managed as the prime tourist attraction of the region. Returning to north Queensland, the future of the last remnants of the primary rain forests seems to rest on the continued tourist success of areas such as the Daintree. (On arrival at Cairns airport, the

Table 4.3 Tourism impact

Accommodation	Fiji	Tonga
Hotels	70	3
Rooms	4,000	200
Motels	20	6
Rooms	350	60
Guesthouses	20	24
Rooms	400	300
Average annual turnover/enterprise ($000)	500	45
No. of units foreign owned	40	–
No. employed	3,500	320
Tourism and travel		
No. of enterprises	45	7
No. employed (full-time)	800	60
Tourist shopping		
No. of enterprises	170	10
No. employed (full-time)	1,000	25
Handicrafts		
No. of enterprises	600	120
No. employed	1,000	150
Tourism as % of GDP	14	6
Gross income from tourism (US$ million)	110	4.5
Leakage (% of gross income)	65	8

first question the immigration officer asked this author was: 'G'day. You'll be going to the Daintree then?') The use of recreation/tourism as a conservation tool requires very delicate manipulation, with careful evaluation of carrying capacities and the fragility/resilience of the ecosystems concerned. Similar dilemmas face the managers of historical and cultural resources; e.g. are tourists sustaining or destroying the Parthenon, or Venice or Macchu Picchu?

Where does this leave the debate? First, it is clear that if the changing conceptualization of the work–leisure relationship spreads to an increasing proportion of the world's population, more and more people are likely to travel for pleasure. They may travel in greater numbers to existing destinations or spread themselves across increasing numbers of destinations. In either case, the economic, social and environmental benefits and costs outlined above are likely to become more intense. Yet growing demand and prospects of profits seem to be almost inexorable forces. For instance, in Thailand, the disbenefits of tourism are fully evident in resorts such as Pattaya and Phuket, but despite expressed government concern, the explosion of mass tourism is rolling southwards. The island of Koh Samui, 32 km off the south coast of Thailand, has long been a popular haunt of adventurers and backpackers. Since the mid-1980s, however, increasing numbers of tourists, disillusioned with the noise and crowds of Pattaya, have been taking the 45-minute flight to Surat Thani and the 90-minute ferry to the island. In 1987, there were 187,000 arrivals, of whom 120,000 were Thais. By 1991 350,000 are expected, and by the year 2000 the totals may exceed 1 million, with the great majority being non-Thais. A Thailand research institute survey

> found that poor water distribution throughout the island was already causing a water shortage in the dry season, leading to conflict between local people and tourists. . . . Of particular concern is refuse – tourists generate 2.5 litres a day per person. Only 25% of the rubbish on Koh Samui is being collected and disposed of. The study also found that local residents did not look kindly on the invasion of their paradise and that only 20% actually benefited in terms of employment or extra income. Although it is not easily measured, moral deterioration is the islanders' growing concern, and they point to higher costs of living, gambling and more drug addiction.
>
> (*Tourism Concern* 2, spring 1990: 9)

The researchers recommend a year-round average of 9,500 tourists a day and a maximum at-one-time capacity of 14,200, a figure they fear will be reached by 1994. Already, several major hotel groups are negotiating for sites and there seems little likelihood that the Thai government will be able to prevent the explosion.

It is clear that there needs to be a changed relationship and changed thinking between the three main components of the tourism realm: the tourists, the tourism industry and the host societies. Increasing emphasis is being placed upon the concept of the 'responsible tourist', one who approaches the visited environment and society with greater awareness and sensitivity. Tour operators are showing signs of enhanced awareness of both the market needs and the need to sustain the quality of the destination and the welcome given by the hosts. The hosts themselves are becoming increasingly sophisticated and organized in controlling the development of tourism more on their own terms for sustained yield. They realize that tourism is a fashion industry, with a life cycle rather like the mining industry. The nightmare of any country, region or resort committed to tourism is that the tourists will move on to fresh destinations, as images and fashions change. They can either try to modify their attractions to accommodate such trends, or plan to use the income from tourism to diversify their economies, and hence to lessen their dependence upon tourism. Examples of the problems can be seen in traditional British resorts which have sought to attract the 'business' trade through conference centres, or indoor leisure centres to counteract the unreliable British climate. The Spanish 'costas' are seeking to recoup on losses of western European tourists by attracting 'less sophisticated' holiday makers from eastern Europe. In the developing world, a crucial key seems to be to maintain control and autonomy in handling the scale and character of tourism and evaluating the role it is to play in development policies.

5 The dumping of radioactive waste in the deep ocean: Scientific advice and ideological persuasion

F. G. T. Holliday

INTRODUCTION

The advanced industrial society of the UK produces significant quantities of nuclear waste; that waste has to be stored or disposed of safely. Radioactive waste is unique in only some respects and it forms part of a much larger waste-disposal problem; other wastes, e.g. that contaminated with some heavy metals, present problems just as challenging. Radioactive wastes arise from many sources: from the nuclear power and defence industries, hospitals, universities and other research laboratories (Figure 5.1). The waste has been classified, according to its level of radioactivity, into 'high level', 'intermediate level' and 'low level'; only intermediate- and low-level waste has been purposely and openly dumped in the deep oceans. In the years 1950 to 1963 waste was dumped by the UK in the Hurd Deep in the English Channel. That site had been used for many years as a place to dump surplus wartime material such as unstable ammunition.

From 1949 the dumping of radioactive waste in the north-east Atlantic was started by the UK, and later other European nations such as Belgium, France, the Netherlands, West Germany, Italy, Sweden and Switzerland also took part. Thus for more than thirty years the UK had been ocean dumping; it had also been disposing of low-level waste into coastal waters, e.g. at Sellafield. Those disposal practices had attracted considerable criticism, not least from the Greenpeace organization.

The Radioactive Waste Management Advisory Committee, which has the statutory duty to advise the government on waste disposal, had permitted ocean disposal to go ahead without the assurance that it was indeed the best disposal option. Also, that committee of technical experts did not take account of the importance of the ideological persuasion of a substantial section of the national and international community. By ideological persuasion I mean the belief by people of the 'rightness' of a situation; it is a matter of conviction and it is hard, if not impossible, to justify or alter such beliefs by means of a technical argument. Ideology is concerned with the morality of a situation; technical advice is

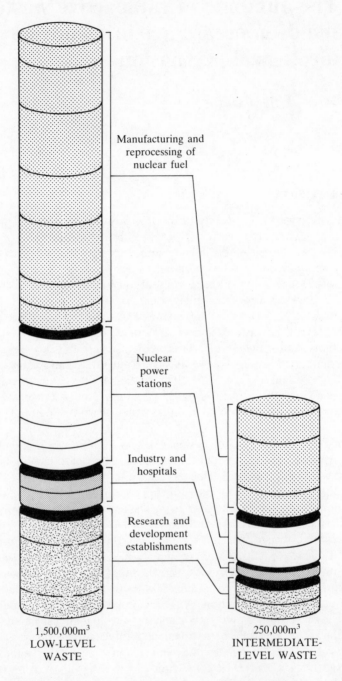

Manufacturing and
reprocessing of
nuclear fuel

Nuclear
power
stations

Industry and
hospitals

Research and
development
establishments

1,500,000m³
LOW-LEVEL
WASTE

250,000m³
INTERMEDIATE-
LEVEL WASTE

Figure 5.1 Wastes projected to arise to 2030 (excluding military sources)
Source: NIREX 1987: 7.

concerned with empirical solutions to problems. Once an ideological objective has been accepted, technical advice can assist in finding the best way to attain the objective. Clearly, a substantial body of national and international opinion had not accepted the 'rightness' of ocean dumping.

It was against that background that events came to a critical point in 1982.

THE ROLES OF GREENPEACE AND THE NATIONAL UNION OF SEAMEN

In August 1982 the radioactive waste dumping ship *Gem* was shadowed by the Greenpeace ship *Sirius*. The intention of *Sirius* was to disrupt and, if possible, to prevent the dumping operation. In the face of that objective a legal injunction not to interfere with the work of *Gem* had been taken out. The law does not command uniform inherent respect, and enforcement is often necessary. An early lesson was that legal injunctions are difficult to enforce on multinational crews operating in remote international waters.

But much more than the law was involved. Members of Greenpeace left the *Sirius* on small inflatable boats; some boarded the *Gem*; others placed their boats alongside, where they were in considerable danger from the falling drums of waste and from accidental collision. All those events were recorded by the television cameras and photographs of the press. The action achieved widespread publicity and caused anxiety and embarrassment. For the crew of *Gem* it was a distressing and unsettling experience. Except in time of war seamen do not knowingly put the safety of other seamen at risk; there must have been considerable tension on board *Gem*. The National Union of Seamen was led by those events to consider the role of its members in the dumping exercise.

The dumping of waste in international waters is influenced by an international 'Convention on the Prevention of Marine Pollution by Dumping Wastes and Other Matter'. Despite its official title, but in accord with its unofficial one (the London Dumping Convention), it works in a way that permits the dumping of radioactive waste and other matter into the oceans.

The convention met in February 1983 and considered a proposal to prohibit the dumping of radioactive waste. That proposal was defeated, but a resolution was passed calling for a voluntary suspension until the completion of a scientific review. That was, in a sense, an 'ideological' point, emphasized by the fact that a vote was needed. Votes at the convention are rare.

The UK delegation did not accept the validity of the resolution on voluntary suspension since it was not based on scientific evidence; however, the UK favoured the scientific review and agreed to accept its findings, and, if thus guided, it would then cease dumping at sea.

The National Union of Seamen followed the proceedings closely and was dismayed by the UK's rejection of the voluntary suspension; the union's general secretary was sympathetic to Greenpeace and held the conviction that ocean dumping was wrong. The NUS persuaded other transport unions to join it in refusing to implement the July 1983 dumping operation, which was called off. The Trades Union Congress met in September 1983 and endorsed the transport unions' action, although not without reservations by some other unions, for example those representing the power workers, whose jobs were closely linked to processes that produced radioactive waste.

There was thus an impasse; waste was continually being produced by processes some of which were ideologically acceptable (e.g. hospital treatment) and some less so (e.g. from the Ministry of Defence). It says a great deal about the complicated nature of the ideological debate that a Conservative government should join with the Trades Union Congress in establishing an independent committee 'to review the scientific evidence, including environmental implications, relevant to the safety of disposal of radioactive waste at the designated North Atlantic site'. The word 'independent' was a clear signal to all interested parties that a 'clean sheet' approach was to be taken.

THE TERMS OF REFERENCE AND MEMBERSHIP

The committee was required to submit its report to the representatives of three interested parties: two secretaries of state (Environment [DoE] and Agriculture, Fisheries and Food [MAFF]) and the general secretary of the Trades Union Congress. It was only after an initial debate that one of those interested departments (MAFF) was persuaded that it would be inappropriate for it to provide the technical secretariat to an independent committee. That technical secretariat was eventually supplied by the Natural Environment Research Council's Institute of Oceanographic Sciences, while the DoE continued to supply the meetings secretary; such a blend proved invaluable. The terms of reference were carefully arrived at, the most important point being the inclusion of the words 'environmental implications'. Behind those words lay the concept that concern for the environment was derived only in part from scientific evidence; other evidence was necessary that took account of ideological conviction. That case was argued cogently by one of the review members and supported to an extent by the others. The scientific expertise of the committee members covered (in summary) marine biology, marine geology and human ecology. The technical secretariat consisted of professional oceanographers. The group had considerable collective experience of public policy formulation and the machinery of government.

SOME TECHNICAL FEATURES

The sea is naturally radioactive, albeit at a low level. It contains short-lived and long-lived radionuclides. That low-level natural radioactivity has been increased, by less than 1 per cent, by human activity through nuclear weapons testing and the sea disposal of waste. Whilst all radio-activity is potentially damaging to living tissues the extent of the damage depends on the nature of the radiation, the intensity of it, the length of time exposed to it and so on. A principle concern about radionuclides in the sea is that they will contaminate those organisms that are used by people as food. Other concerns include the contamination with radio-active debris of coastlines and water used by people for work or recreation. Unless the damage is at an extreme level it is not easy to detect the harm caused to an individual or a population by radiation. For an individual, the risks from exposure to radiation may be very much less than other risks to health and life, e.g. from smoking or the risk of traffic accident. However, when the small risks to any individual are recalculated to apply to a population exposed over many generations, the changes of scale involved may profoundly alter the perception of the problem. Any damage done to the genetic material passed from parent to offspring is especially significant; it is exposed over many generations and is redistributed according to mating patterns.

In assessing the risks it was important to examine the barriers between the radioactive waste and the seawater, the rate of release of radioactivity when those barriers eventually broke down, the dispersion from the dump site of contaminated seawater and the possible pathways back to man, whether in food organisms or otherwise, of the radionuclides.

The original waste materials included solidified sludges, metallic objects (e.g. tools, pipes, needles), crushed laboratory glassware, rubber gloves and other protective clothing, resins used for chemical processing, incinerator ash and various plastic containers. Special attention was given to the plastic waste since it consisted of up to 10 per cent by weight of the whole and, because it is buoyant, can reach surface waters directly and relatively quickly. In the belief that thereby it could be made to comply with the London Dumping Convention, such plastic waste was shredded; that shredding served to increase the exposed surface area and multiplied the distribution problem.

The waste was packaged into concrete-lined steel drums; it was estimated that the steel had a life of 30 to 40 years and the concrete lining 300 years or longer. The drums were dumped over the side as the vessel steamed slowly over the dump site. The site itself (see Figure 5.2) is large, about the size of Cornwall. It is in the foothills of the Mid-Atlantic Ridge and the average depth of the water is 4,400 m (about 2.5 miles). It is not easy to study the living organisms, the sediments or the water movements at such depths. On the site the ocean floor is covered with a layer of sediment more than 300 m thick. Animal life is present, including large numbers of nematode worms, a scavenging amphipod shrimp and a few fish (mainly rat-tails). There are almost certainly significant populations of micro-organisms representing the base of many food chains and biochemical cycles.

Water movement is slow (1 or 2 cm per second) and the sediments relatively undisturbed, although almost certainly there will be enough movement to keep a layer of fine particles in both suspension and slow movement just above the surface of the thick sediment layer. The water at that depth is cold and dense and thus forms an unmixed, slow-moving bottom layer.

Thus the drums are scattered over a large area, into sediment, at great depths where pressures are high and total darkness prevails except for any biological light. Retrieval of the drums would be an immensely difficult if not impossible task, especially when corrosion has progressed.

WHAT HAPPENS WHEN THE WASTE ESCAPES?

It is not possible to be certain about the processes that would follow the escape of the wastes. However, some intelligent and informed guesswork suggests that the radionuclides that dissolve and diffuse out of the canisters will at first remain in the water layer close to the sediment surface. They will then gradually be spread and diluted by the slowly moving water and by further diffusion. Although the analogy is very far from perfect, the gradual dispersion and spread of a plume of smoke from an extinguished candle in a vast, still cathedral might help in visualizing the situation. It is estimated that the dissolved material would take about 20 years to circulate around the deep water of the site and 200 years before it came within 2,000 m (about a mile) of the surface. But certain other active processes are also occurring: first, radioactive decay will have already reduced the activity of the material in the canisters, and will continue to do so on release. Materials such as iron–55 (half-life 2.7 years) will to all intents and purposes have lost their activity long before release. Others, e.g. plutonium–239, with a half-life of 24,113 years, will far outlast the drums. Also, the radioactive materials will be absorbed by the particles of sediment floating above the sea-bed; some of those particles will be ingested by living organisms which in turn have varying powers of movement. Those living organisms will be part of a food chain which could, in certain circumstances, end in man.

It might be thought that studying such a sequence of events would be a relatively straightforward exercise, a matter of taking samples of water, bottom sediments and living organisms from the vicinity of the dump site and analysing them for radioactivity. However, that is not the case. The steel and concrete packaging delays the release; some substances leak out more rapidly than others; the natural background radiation will

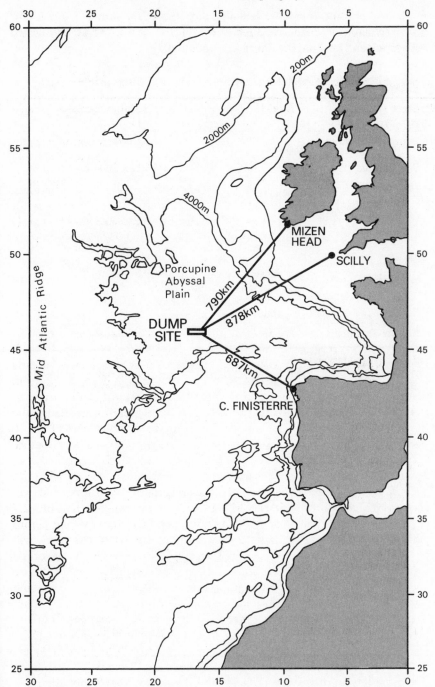

Figure 5.2 Location of the present north-east Atlantic dump site
Source: Holliday 1984: 24.

mask any chronic low-level leakage. For those and other reasons it is not practicable to measure directly the impact of the dispersion of the waste; instead the impact must be modelled.

Figure 5.3 Modelling framework used in radiological assessment

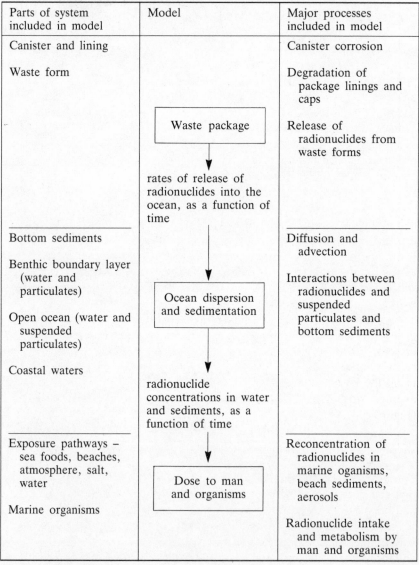

Parts of system included in model	Model	Major processes included in model
Canister and lining		Canister corrosion
Waste form		Degradation of package linings and caps
	Waste package	Release of radionuclides from waste forms
	rates of release of radionuclides into the ocean, as a function of time	
Bottom sediments		Diffusion and advection
Benthic boundary layer (water and particulates)		Interactions between radionuclides and suspended particulates and bottom sediments
Open ocean (water and suspended particulates)	Ocean dispersion and sedimentation	
Coastal waters		
	radionuclide concentrations in water and sediments, as a function of time	
Exposure pathways – sea foods, beaches, atmosphere, salt, water	Dose to man and organisms	Reconcentration of radionuclides in marine oganisms, beach sediments, aerosols
Marine organisms		Radionuclide intake and metabolism by man and organisms

Source: Holliday 1984: 32.

Modelling in some form is used by most people to help predict situations impossible to measure directly. For example, when a long journey by car needs to be made, and a time of arrival has to be indicated, a 'model' of the journey is made. A route from start to finish is worked out; the mileage is calculated; an average road speed is assumed taking account of known roadworks; allowance is made for stops for refuelling and meals; and at the end of all that an approximate time of arrival can be estimated. The costs of the journey can also be calculated; the miles per gallon achieved by the car, the costs of fuel per gallon and the total mileage are combined in the calculation. Although the models are an estimate, at least parts of them can be tested, e.g. the miles-per-gallon of the car. Other parts of the model can have a range of likely variation put upon them, e.g. average speeds of the various types of road. The answers might be approximate, but they are close enough to be useful and some possible sources of error can be identified and estimated, e.g. a mechanical breakdown or a puncture.

Modelling is then a commonplace aid to dealing wih uncertainty and practical difficulty in our everyday lives. Of course there is a vast difference in scale and complexity between the example given and the modelling of events at the dump site, but the principles are to all intents the same. The waste-dispersion model divides into three parts: a waste package model to estimate the release of the waste from the containers; then a set of dispersion models which estimate the dispersion of the released waste in the water, sediments and food chain; finally, a set of models to assess the doses to man via pathways through the food chain or material washed up onto the beaches. Figure 5.3 summarizes the modelling framework.

A preliminary set of results from the models indicates that the highest doses of radiation will come from the consumption of molluscs, crustacea or seaweed harvested from the Arctic Ocean during the period 85 to 95 years after the start of dumping. The maximum dose to an individual, using the worst predictions, is 0.8 μsv, i.e. 0.04 per cent of the annual average dose from UK natural background radiation to individuals (Figure 5.4).

THE BEST PRACTICABLE ENVIRONMENTAL OPTION (BPEO)

The crucial recommendation of the independent committee's report was that sea dumping of radioactive waste should cease until further studies had been made, especially a study of all the disposal and storage options so as to establish which of them represented the Best Practicable Environmental Option, i.e. the 'best' option based on occupational and environmental cost and social considerations. The report was published in November 1984 (Holliday 1984); the government accepted its recommendations and an assessment of the BPEO for low-level and intermediate-level solid radioactive waste was published in March 1986 (DoE 1986). The five options considered were: sea disposal; near-surface burial in

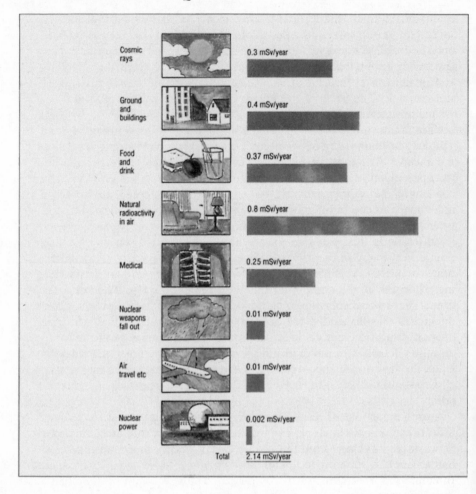

Figure 5.4 Average sources of radiation in Great Britain (millisieverts/year)
Source: NIREX 1987: 11.

either simple trenches or concrete-lined trenches on land; burial in deep cavities on land; disposal down boreholes under the coastal sea-bed; and long-term monitored storage on land.

The study was a thorough one; it had to be, since over 70 different kinds of waste were identified from more than 400 separate 'streams' of origin. An estimate was made of the amount of waste that would arise up to the year 2030, based on a moderate growth in nuclear power. The study may therefore have slightly overestimated the size of the disposal problem, but a 'no growth, no fuel reprocessing' option was also evaluated without altering the conclusions. Regrettably the 1986 DoE study (as in the previous 1984 Holliday Report) was unable to take open account of Ministry of Defence activities, although assurances were apparently given that such wastes were similar in kind to other wastes and produced in quantities that would not affect the conclusions of the study.

One assumption continued to be made in the study that is unsound both in scientific terms and in terms of the London Dumping Convention. The assumption was that persistent buoyant plastics, contaminated with radioactive waste, would be rendered suitable for sea disposal if they were shredded and grouted in concrete. Such an assumption had been challenged earlier in the Holliday Report and the challenge had presumably been accepted by the government. The BPEO study gave little weight to either the Holliday challenge or the fact that it had been, at least implicitly, accepted by government. The London Dumping Convention is very specific in prohibiting the dumping of persistent plastic and other materials which may float. It seems an obvious point that such materials would, over time, inevitably be freed from the restraint of the drums and concrete grouting; they would then become the largest, most direct and least diluted route back to the surface waters. The shredding of such material exposes an even larger surface area of contamination and thus exacerbates any problem. Whilst the risk to individuals is negligible, to ignore that route to the surface, whilst making extreme and expensive efforts to calculate all other routes, is paradoxical, to put it mildly.

In the event, as will be seen, that flaw in the study is likely to have little practical consequence. Having identified the quality and quantities of waste and selected the five basic disposal options available, the study had to apply a set of criteria to the five options to determine 'the best'. In comparing the options it became clear that no one option was better than another in all respects. Some attempt was needed to rank the options, so a 'weighting' system was adopted. Four sets of weightings were chosen and applied to the five disposal options; the same weight was given in each set to reducing individual risk, but the sets varied in other respects. In summary:

Set I was chosen to reflect reduction in costs as a highly desirable feature. It also took account of radiation risk to workers and short-term radiation exposure to the public. No value was given to radiation risk from widespread dispersion of radioactivity in the environment after 1,000 years.

Set II was less concerned with costs but had the same desire to reduce risks to individuals and collective doses. In this second set no discrimination is made between local, regional and global impacts.

Set III reflected a concern to minimize local impacts, whether to workforce or resident population. Little weight was given to reducing costs and a high weight was given to reducing risks and doses from, for example, accidents at stores.

Set IV in some ways represented a fundamental change in thinking. It represented an option with the least environmental impact, whatever the cost; there was a very low weight given to costs and a high weight given to minimizing far-future collective doses. Greater weight was given to avoiding even low levels of contamination that would persist for long periods over large geographical areas.

It was argued that Set I favoured the nuclear power industry; Set II was 'neutral'; Set III favoured local populations; and Set IV favoured 'environmentalists'. The outcome of the study was that all of the disposal options were considered to be both practicable and safe for appropriate dumping and that all four weighting sets showed a preference for some use of sea disposal. However, by the criteria and standards of the study it was concluded that land-based, near-surface disposal is the Best Practicable Environmental Option for over 80 per cent of all the wastes considered. Sea disposal was the BPEO in all four sets of weightings for only 15 per cent, including certain Magnox ion-exchange resins and tritiated wastes (so long as those wastes were free of C14). The sea-disposal option had, of course, one advantage over all the other options considered; it already existed and had an infrastructure of packaging, transport, site and site-specific assessment in place. However, the site lay outside the UK national boundary and thus outside the autonomy of the state. The option depended on international opinion and international acceptability; such opinion was divided and vocally critical. The study concluded that, should sea disposal be totally abandoned, the increases in the use of land-disposal facilities would be so small as to have a negligible effect on their radiological and social impacts. On the other hand, to continue to use sea disposal for the small amount of waste for which it was the best option would imply a commitment to the costs and international political complications which had brought about the original crisis.

THE ROLE OF NIREX

The Nuclear Industry Radioactive Executive (NIREX) was established in 1982 in order that radioactive waste of all levels of activity should be disposed of to high standards of safety of a minimum period of storage. It is a company with shares held by other companies concerned with waste production. The government holds a special share. Since 1983, NIREX has considered a number of 'on-land disposal sites'. Such sites need to be chosen according to well-defined geological and hydrological features. In 1986 the government announced that all intermediate-level waste should be disposed into deep depositories and in 1987 it announced that low-level waste should be similarly disposed of. In 1987 NIREX

Table 5.1 The advantages and disadvantages of three concepts of radioactive waste disposal

Concept	Advantages	Disadvantages
Under land	Feasibility of excavations proven by precedent of mining Relatively simple ground investigations Single handling of waste at surface Less expensive concept, with confidence in cost estimates International consensus of concept Easy access	Under someone's 'back yard'
Under the sea-bed accessed from the coast	Feasibility of excavation proven by precedent of mining Single handling of waste at surface Simple waste emplacement Concept comparable to under land International consensus of concept Easy access	High cost of ground investigations Possible international political ramifications Legally more complex
Under the sea-bed accessed at sea	Away from anybody's 'back yard' Low groundwater flow likely at depth under sea	Concept not proven by precedent High cost of geological investigations Likely to be most expensive concept with least confidence in estimates Double handling of waste, at port and offshore Sea transport subject to weather Platform provides a limited operating work area Possible international political ramifications Legally more complex

Source: NIREX 1987: 21.

published a discussion document entitled *The Way Forward* in which the various issues involved in choosing the site for a deep depository are outlined. The purpose of the publication was to inform and seek contributions to the debate. Such an attitude is to be welcomed, although the outcome is predictable and there is strong opposition from the local populace to any area or site indicated as a possible disposal area. NIREX tabulated the advantages and disadvantages of various broad categories of disposal area (see Table 5.1) and once again the sea-bed was seen as having the advantage of not being within anyone's 'back yard' (a reference to the so called NIMBY – 'not in my back yard' – phenomenon). However, in this study disposal would be beneath the sea-bed and any direct contact between the waste and seawater is not envisaged.

CONCLUSIONS

At the time of writing, the future of radioactive waste disposal remains uncertain and the most likely outcome is that the waste will remain within the boundaries of government-owned land already dedicated to the nuclear industry, with some form of compensation paid to the local community. The publication by NIREX of the document *Going Forward* (1989) and the drilling now taking place at Sellafield and under consideration at Dounreay add support to this view.

Recent evidence (for example, a letter received by the author in October 1989 from the UK Atomic Energy Authority following publication of its 1988 Annual Report) suggested that the government does not expect to resume deep-sea dumping of wastes, but it has reserved the option to dispose into the deep sea certain large components arising from the decommissioning of large facilities. The UKAEA no longer regards it as appropriate to make provisions for the costs of deep-sea disposal. The decommissioning of out-of-date power stations and nuclear-powered submarines will test the government's intention.

Whether it will avail itself of the deep-sea option for such large items of waste, and how it would justify such disposal in BPEO terms, awaits to be seen.

6 The ocean environment: Marine development, problems and pollution

Philip Neal

About 75 per cent of the surface of Planet Earth is covered by the oceans and their contiguous seas, in all some 140 million square miles down to depths of 7 miles. 'Planet Ocean' might well be seen as a name more apt for our part of the solar system (Figure 6.1). So vast, yet so far removed from the day-to-day lives of the majority of people of the world, the marine environment is viewed by most of mankind as everlasting and something so immense that it is immutable. Its waters bathe the entire planet and its artery-like currents influence the countries which lie within their 'hinterland'; its freshwater extensions reach into the heart of the human habitat. Despite the ebb and flow of the tide and the exaggerations brought about by storm-induced waves, the oceans and seas remain within well-defined limits.

From their waters we collect food and extract salt, from their beds exploit minerals and pump gas or oil, on their surface transport goods and people, and from their tides, waves and currents create power. Into their waters we dump rubbish, pour polluted water and with careless lack of concern perpetrate one accident after another, thereby spilling poisons into the depths and crude oil onto the surface. With complete indifference to the damage done, perhaps because it cannot be seen, we disturb the sea-bed with our mining activities and sink the materials too dangerous for us to contain on land. Yet all of the while we rely on the oceans to fuel the water cycle and to act as a levelling mechanism for the extremes of temperature variation.

There are danger signs that all is not well; the constant abuse of this combined larder, dustbin and heat conditioning system is beginnning to be seen. The evidence is plentiful: creatures with deformities, sea birds dead or dying, coasts invaded by algal blooms, fish nets empty, floods, islands overrun by the sea, beaches unable to pass the minimum of health standards laid down by monitoring bodies, common marine animals of but a few decades ago no longer seen, and complete areas of water devoid of life.

Overall the oceans still appear to tolerate us – but for how much

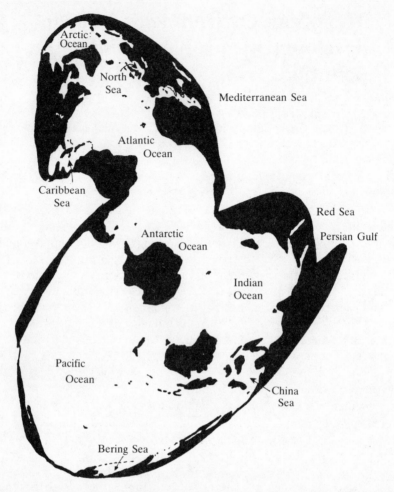

Figure 6.1 Planet Ocean (Spilhaus whole ocean map): by splitting the world along land axes the contiguous nature of the ocean world and its areal dominance over the land become obvious

longer? In the words of the scientist whose name is for many synonymous with that of the ocean world:

> The health of the global water system rooted in the ocean is vital to the future welfare of our planet, and is of particular concern to me as an ocean explorer. The future needs of society will be well served, however, only if we change our short-term mentality and often arrogant indifference to the results of our actions and focus on long-term considerations and a sound attitude in the use of all our resources.
>
> (Cousteau 1981: xix)

This chapter cannot give more than a brief overview of the marine environment and can only touch on some aspects of sustainable development. It will concentrate on the damage being done to the waters of the vast oceans and seas by the excess discharge of nutrients; this will be related to the disposal of sewage sludge. Industrial wastes in general and oil pollution in particular will be discussed, as will the rubbish which collects on the shoreline. Commercial fishing is an important aspect of the use made of marine life, while coastal disturbance and global warming are both environmental issues on a world scale which affect the oceans and seas. Mining the sea-bed and electricity production are in their infancy but may present many problems technological and political, not least of which is the lack of international agreement. Finally marine legislation will be considered briefly.

Life abounds in the oceans, particularly in the waters over the continental shelf at depths to which sunlight penetrates. Vegetable and animal plankton in abundance provides food for larger marine creatures, underwater forests of kelp flourish, and the benthic (bottom-dwelling) organisms consume the detritus that reaches the sea-bed. Most rely on seawater nutrients, which are chemicals, both elements and compounds, essential to their growth. Nutrients are no pollutants but when excessive quantities are released into the oceans they often have an adverse effect.

Quantities of nitrogen (N) and phosphate (P), which react positively on the marine life of the waters of the organically rich ocean rim, have increased over the past fifty years. This situation arises from the discharge of sewage (N and P), agrochemicals (N and P) and detergents (P) into rivers and sea.

As grass is to land so phytoplankton is to water. Both lie at the bottom of the food chain; marine life depends on the grazing of the vegetable plankton by the small creatures of the zooplankton. When the marine system is in balance the production of the phytoplankton is matched by the consumption of the zooplankton. Limit the vegetable matter and the zooplankton will be limited too; consequently fish and higher forms of life will be affected adversely. Where extra nutrients lead to an overproduction of phytoplankton the excess begins to rot and in so doing removes oxygen from the water. This, in turn, leads to damage to the marine ecosystem. Eutrophication is the term often used to describe this state of affairs. This is an unfortunate use of the word for in a strict etymological definition it means an environment with sufficient nutrients present to sustain growth. Hypertrophic (Gerlach 1988) is a better term to use to describe a situation where excess nutrients are placing stress on an ecosystem, in some cases so much that to all intents and purposes the sea area is 'dead'.

The North Sea has been investigated in considerable detail and it has been concluded that 70 to 80 per cent of the N and P in the waters off the coast of Denmark, West Germany and the Netherlands (mainly the Wadden Sea) is of anthropogenic origin. This is the area of the North

Sea most threatened by hypertrophication (Nelissen and Stefels 1988). Marine algae can produce dimethylsulphoxide (DMS). Measurements suggest that emissions produced by North Sea algae in spring and summer are equivalent to 25 per cent of the man-made sulphur in the atmosphere.

DUMPING AND ACCIDENTAL DISCHARGE

The deliberate discharge into the sea of waste or other material is known as 'dumping'. Materials dumped will be examined under the following headings:

- Sewage sludge.
- Industrial waste.
- Oil pollution (including accidental discharge).
- Rubbish (domestic, industrial and shipping).
- Dredged material.
- Ocean incineration.

These originate from domestic, industrial, agricultural and other sources. Materials which are dumped or are accidentally discharged include the following:

- *Domestic*: sewage, rubbish, silt and other debris from roads.
- *Industrial*: sewage effluent, waste (biodegradable), waste (toxic), waste (persistent), rubbish from shipping, oil, dredged material, radioactive materials (see Holliday, Chapter 5 in this book), inert solids, inciner-ation-at-sea products, thermal discharge.
- *Agricultural*: organic matter, fertilizers, biocides (fungicides, pestici-des, herbicides).
- *Other*: munitions, medical materials, lost cargo from ships (hazardous and non-hazardous).

As soon as these substances and materials enter the water they may undergo biological, physical or chemical change which may distribute any contaminants and the products of any interaction between the water sediments and the atmosphere. They may be transported throughout the oceans by the currents, which means that any prognostication of their environmental effect is very complex. The biological effects include the bioaccumulation of trace metals; the physical effects include the smother-ing of benthic organisms; and the chemical effects include the alteration of the acidity (pH) of the water. Certain materials can build up to levels which are toxic to organic matter – for instance, mercury and pesticides. These can bioaccumulate within organisms and biomagnify up the food chain. This leads to high concentrations in the larger creatures with a possible adverse effect on people. The most infamous case is probably that of the industrial discharge of mercury into the waters of Minamata Bay between the years 1953 and 1960.

Other materials may not be toxic but they can interfere with biological activities. For example, the masking of the pheromones by falling sediment results in a diminution of the sexual attraction between organisms and thus a reduction in breeding success.

In considering dumped material some account needs to be taken, not of the precise quantity entering the seawater, but of the bio-availability of the contaminants. For instance, trace metals within sewage sludge, such as the mercury, are far more likely to be available to organisms than those contained within fly ash or colliery waste.

Another aspect of marine pollution is a consideration of the pathways by which substances enter the oceans or seas. Some enter via rivers, some by direct dumping from outfall pipes or ships and other material from the atmosphere as a result of aerial discharge from power stations and industry. The environmental effects result from a summation of all inputs to the sea.

Contaminants will be spread between the water itself, the sediments, the atmopshere and the living biota. They will be transported by winds and currents around the world so that traces of DDT are found in the penguins of Antarctica, thousands of miles from any possible source.

A thin oily film, called the *lipid microlayer*, is present on the surface of the sea. It is as thick as a one penny piece at its maximum. It is the site of high biological activity, the haunt of the organisms that float or swim in surface water. It contains the eggs or larvae of many invertebrates or fish species which feed on the lipids, fatty acids and other dissolved organic matter so abundant in this surface layer. Unfortunately many of the anthropogenic materials, including hydrocarbons, metals and other toxic pollutants and the most deadly of poisons the non-biodegradable dioxin, are found in this micro layer in concentrations higher than in sub-surface water. This is partly due to their deposition from the atmosphere and partly to upward movement through the water column. Several varieties of sea bird skim off this water layer to collect their morsels of food, and together with the rest of the fauna they are liable to adverse effects from the toxins.

Sewage sludge

Sludge is categorized into three states:

- Raw: untreated.
- Digested: sludge that has been fermented until it decomposes.
- Activated: digested sludge which has a second fermentation, is then dried and can be used for dry fertilizer unless it contains trace metals harmful to agriculture.

Raw sewage which has received the minimum of attention, having passed through coarse grids to remove the larger items of solid material, is often considered inaccurately as 'treated sewage'. Such claims are usually made by those defending the status quo of sea dumping via outfall pipes; whatever is claimed, the discharge is still raw sewage.

Many countries of the world discharge sewage into the sea, defending the action by claiming the effectiveness of the remedial treatment by seawater; some even claim that sewage is of benefit to marine life. Both claims may be true given sufficient time for the chemical breakdown to occur and where the discharged amount is limited in quantity. The real motivation is economic, with little attention paid to sustainability or the true environmental costs of minimal sewage treatment; yet proper treatment could turn a problem into an asset as an organic fertilizer for agriculture. Even the strongest protagonist of marine disposal of sewage cannot deny the aesthetic objection to faecal deposits and other macro material defacing the shoreline. In its review of sewage sludge disposal the Oslo Commission (1989, quoted in Clegg and Horsman 1990: 42) reports the UK authorities as saying that 'the coastal waters around the UK have been little affected [by sewage sludge dumping] and that there has nowhere been a significant or unacceptable detriment to the marine resources which these waters support'. It all depends on the definition of 'unacceptable detriment'. In fact what is, or what is not, an 'unacceptable detriment to the marine environment' is fundamental to the environmental debate – no definition has been given!

Industrial waste

The type of industrial waste dumped in the sea will depend on the nature of the processes carried out in the lands local to that sea. Nevertheless most of the industrial economies of the western world are fossil-fuel based, so that the bulk of the solid disposal is of colliery waste and fly ash. The remainder includes liquid slurries, as for example the waste acids from bulk chemical manufacture, pharmaceuticals or specialist chemical manufacture, food, textile and fibre manufacture, the metal finishing industries and caustic soda solutions from petrochemical plants. Bio-degradable wastes such as hydrocarbons from industrial processes, silage material from intensive farming activities and various materials from the manufacture of synthetic fibres and pharmaceuticals demand a large amount of oxygen to break them up. This may cause areas of oxygen depletion known as anoxia. For certain toxic chemicals the marine environment has no assimilative capacity.

Oil pollution

The oceans and seas will always accumulate oil contamination due to the natural seepages from oil-bearing rocks on the sea-bed. Such is the vastness of the marine environment that this can be accommodated without serious damage to the living organisms. This cannot be said for the pollution which has arisen from the oil-based economies of the world. This is not helped by the fact that areas of oil consumption are usually great distances from the places where the oil is produced. This leads to an enormous increase in the possibilities for oil spillages which occur at the production, distribution and refining stages; the majority of them affect the marine habitat.

Pollution by petroleum hydrocarbons in estuaries and coastal areas, as a result of discharge from oil refineries, occurs particularly in parts of the world where regulatory restrictions are fewer or more leniently enforced.

Sea-based oil rigs can add to the problem of oil spillage in two ways. The first is the use of oil-based cutting muds which lubricate the drill heads and so escape ino the surrounding sea. This is not a major problem, and with the introduction of lower-toxicity mineral oils in the formulation of cutting muds, pollution from this source should decrease over the years. The second is for accidental discharge, from the minor leaks which inevitably occur with the complicated pipework of extraction and distribution, to the major escapes of oil rig catastrophe such as have been experienced by Piper Alpha (1988) off the east coast of Scotland, and Ixtoc 1 (1979) in the Gulf of Mexico. Yet even here individual circumstances determine the amount of biotic damage. Piper Alpha was a major catastrophe in human terms, with 167 deaths, but the intense incineration effectively removed the crude oil discharge before pollution took place. In the case of Ixtoc 1, 140 million gallons of oil escaped over 295 days before the well was capped in March 1980.

Illegal discharge from ships either cleaning out their oil-carrying tanks or performing routine maintenance on their diesel engines leads to a considerable amount of oil pollution. There needs to be increased inspection of tanker equipment in ports and more surveillance of ships and oil installations at sea to give a greater possibility of detecting illegal discharge. 'Knowledge of possible prosecution is perhaps the most effective way of reducing the incidence of inadequately maintained equipment and human failings, which have been shown to be the cause of spillages'; (Leatherhead 1990: 50).

It is not the general low-level background and contamination which causes the greatest biological damage (Stowe and Underwood 1984) but the devastating spills resulting from tanker wreck or collision. *Torrey Canyon* (1967) and *Exxon Valdez* (1989) are but two of the names now infamous in marine history.

The actual effects of oil pollution are sufficiently well publicized to limit comment here. Some aspects of the cleaning of oil spillages invite attention. Detergents may be spread on the slicks to break up the oil residue and to cause it to sink. More research is needed to investigate the effect on the marine environment of oil falling to the sea-bed; at the present time it appears that clean-up success is measured by the disappearance of the contaminant from sea surface and shore. In many cases the break-up of large slicks has to be left to natural forces; how quickly this takes place depends to a large extent on the physical properties of the oil itself. Whether the oil is 'light' or 'heavy' is of paramount importance in the persistence of the problem.

Rubbish

Unfortunately most shorelines are marked by a zone of rubbish brought there by the action of wave and tide and deposited at high water mark. During the last decade attention has been drawn to the increase in the widespread and continuous inputs of solid wastes, particularly non-biodegradable plastic, into the marine environment. Harmful effects upon marine organisms have been identified worldwide, especially on mammals, sea birds and fish. The major source of this marine litter is considered to be the dumping discharges from ships at sea, although land-based sources have been recognized including rivers and holiday beaches.

The garbage deposited on the shoreline is aesthetically unpleasant and hinders the proper enjoyment of the seashore. It provides health hazards, particularly of injury resulting from broken glass, metal protuberances and other sharp edges. This is bad enough but the dangers to marine life are more alarming, resulting from entanglement or ingestion. Discarded netting accounts for the deaths of many of the larger marine mammals such as seals, whilst fishing line and the four- or six-pack plastic yokes for drink cans take their toll of smaller animals and sea birds. Ingested plastics are hazardous to all creatures, especially sea birds. One survey of the contents of gizzards of sixty-five fulmars found dead on Netherlands beaches in 1983 found that 92 per cent contained plastics, with a mean average of 11.9 plastic objects per bird; one poor creature had ingested 96 plastic fragments (Dixon 1990).

Facilities for the proper disposal of waste are needed at all ports, and increased surveillance is necessary to implement the MARPOL regulations to be discussed later.

Dredged material

Most harbours, ports and estuaries require dredging operations in order to allow free access to shipping. Almost all of this spoil is dumped at sea. Much of this material is contaminated with heavy metals which have been deposited from their industrial or domestic sources, and DDT, dieldrin, dioxin, PCBs (polychlorinated biphenyls) and petroleum hydrocarbons are also included in this devil's cocktail. The dredging process is in effect a method of transporting pollutants from rivers into the open sea, where it is deposited on the sea-bed to the detriment of the benthic species. Masking of the pheromones has already been mentioned.

Whereas the pollutants become relatively stable in their estuarine environment, their disturbance and in particular their second depositional fall to the sea-bed make the release of trace elements increasingly available to the marine organisms. It is apparent that much sewage sludge which had come to rest in estuaries is disturbed and 'reactivated' by dredging. There is evidence that some of the dredged material returns to estuarine areas due to the coastal currents and the effects of the tides.

Ocean incineration

The waste which is generally incinerated at sea is largely made up of intractable inorganic chlorine compounds. These are extremely toxic and must be destroyed at sufficiently high temperatures. Most investigators agree that maximum efficiency is unlikely all of the while; partial burning leads to the release of fragments of the original material which will recombine away from the smokestack. Another cocktail of highly toxic chemicals is formed, and recent evidence suggests that the effects of complex organic compounds such as these may be highly additive (Pederson 1988).

At the London Dumping Convention in 1988 it was agreed to ban ocean incineration globally by 1994. This does not include the incineration of rubbish or oil waste.

COMMERCIAL FISHING

Even those people of the developed world remote from the sea are aware of its existence as a source of food. The improvements in refrigeration in both distribution and domestic storage have meant that fish is a staple food in most diets. Despite the obvious danger to fish stocks of overfishing there is serious opposition to control procedures, in particular the setting of restrictive quotas and the size of the net mesh in use. Industrial fishing (where the quarry is marine life not destined for the human palate but turned into animal feed or fertilizer) can lead to problems. The sand eels of the northern coasts of Britain are a case in point. Their increased exploitation is having the inevitable consequences on the higher forms of fish which rely on the sand eels as a source of food and on sea bird colonies too.

The traditional methods of catching fish are still by drift net (gill net) and seine net for surface and mid-water species and by trawl for bottom dwellers and some mid-water fish. The traditional beam trawl has a beam mounted at the entrance to the trawl net to keep the mouth open. Heavy tickler chains drag along the bottom and force the fish into the nets. The complete system weighs about 5,000 kg nowadays, contrasting with the lighter trawls of twenty and more years ago. These lighter trawls disturbed the bottom sediment to a depth of only one or two cm; now the heavier equipment churns up the sea-bed to more than 6 cm. The organisms which live in this top stratum provide the base of the food chain for the commerical fish. Overfishing means that the same area may be trawled between three and five times a year.

In addition to this disturbance and its effect on the fish population, the whole of commercial fishing is very wasteful. All fish that are caught die; for every 1 kg of useful fish landed, between 2 and 4 kg of unwanted dead fish are returned to the sea. This does not take into account the many other species which are trawled in, crabs for example, which may or may not survive the ordeal. Trawling also takes its toll of sea mammals – it is estimated that Danish trawlers alone account for the deaths of almost 4,000 dolphins, seals and other large creatures every year.

Drift net fishing has also been a traditional method but, unfortunately, fine microfilament nylon netting has been introduced, seemingly invisible to marine life. In the Pacific Ocean, Japanese, Korean and Taiwanese fishing fleets are using drift netting over 100 miles in length which sweeps up everything in its path. This includes sea birds as they dive for food, and seals, turtles, small whales and dolphins as they struggle to surface for air. These three countries capture about 40,000 tonnes of tuna annually, which is about twice the amount which fishing experts consider should be the maximum catch to maintain tuna stocks. The amount of other marine life captured and thrown away is not known. The fifteen nations of the Pacific Forum are attempting to ban the use of these gill nets, known to environmentalists as the 'wall of death'.

In the North Atlantic some 20,000 porpoises die every year in the salmon nets. The capture of unwanted species is always a problem and involves consideration of the morality of sacrificing unwanted animal life for the gratification of the human palate.[1] Those who refuse meat on ideological grounds but who include fish in their diet are, hopefully, just ignorant of the facts, rather than hypocritical.

Attempts to control the stocks of commercial fish by quotas have had some success. In the North Sea stocks of herring, mackerel and haddock have shown some recovery since the European Community attempted to control their disastrous decline in stocks. Yet the fishermen are loud in their opposition to any limitation in the size of their catches. The task of policing and enforcing legislation in the wide-open spaces of the oceans is a mammoth one and almost impossible to do effectively. Individual states have taken unilateral action to prevent overfishing. Perhaps the best-known example in Europe is the action of Iceland in the 1970s of creating an exclusion zone around its shores to protect the stocks of cod.

The increase in demand for fish products is the economic motive behind the global increase in the number of fish farms and kelp 'fields'. Of great environmental concern is the effect of the detritus falling from the cages, which may be droppings from the fish themselves or unused feed. The effects on the sea-bed is similar to that of the discharge of sewage, with the extra nutrients bringing about all the ill effects mentioned previously. The close proximity of farmed fish to one another means that the spread of infection and parasites is a problem which has to be tackled, usually by the introduction of chemical products. Nuvan is a good example; it is used to kill parasites, but however careful the 'farmer' is, there is an inevitable escape into the surrounding sea. Here it will continue to poison small organisms and, just as rose sprays tend to kill off both the aphids and the bees, the Nuvan will poison parasites, the larvae of shellfish and the fry of fish species.

Escape is also a problem with salmon. Some will escape into the estuary and meet up with wild salmon. The result may well be interbreeding which, it is feared, will reduce the ability of the salmon to face the rigours of its breeding cycle, involving as it does great journeys into the oceans and exhausting travel back up the rivers to the spawning grounds. Already there is some evidence that escapees from Norwegian salmon farms have appeared in the coastal waters of the north of the UK.

Shellfish are at risk from the use of anti-fouling paints. These paints are applied to the keels of boats to prevent the build-up of the larvae of shellfish on the surface. This discourages the attachment of barnacles and other limpet-type organisms which reduce the efficiency of movement through the water. It is in the nature of these paints to kill organisms which come into contact with them and to deter any settling on the hulls of boats. By the same token they are lethal to any commercial shellfish which come into contact with them. In particular the introduction of paints containing the very toxic tributyl tin (TBT) and the increase in estuarine and coastal marinas have meant a wiping out of some areas of shellfish production.

COASTAL DISTURBANCE

Shellfish production is also adversely affected by the reclamation of coastal zones for whatever purposes, as are other coastal flora and fauna. The removal of mangrove swamp can be extremely deleterious. It is possible to target several marine species in danger of serious depletion, if not extinction, through coastal zone disturbance – turtles are one example. Turtle decline applies to all tourist areas where turtles are indigenous to the holiday beaches.

In an attempt to counter the problem of the exploitation of breeding beaches, and a proper implementation of reserve regulations, the countries around the Mediterranean have set up MEDASSET (Mediterranean Association to Save Turtles). The aim of this association is in opposition to the tourist developers; and storms, other than those of nature, are to be anticipated in the next few years.

GLOBAL WARMING

There appears to be little doubt that atmospheric pollution is leading to global warming through the so-called greenhouse effect. It seems to be inevitable that mean sea level will rise, chiefly because of the expansion of the water due to warming, effectively flooding the lowest-lying areas of land. Estimates vary, but a general consensus of opinion suggests a rise of up to 2 m over the next century. On top of this will be the regular inundation of land presently 5 m above mean sea level, as storm surges and tidal variations probably increase.

The marine situation will be exacerbated by a change in the pattern of violent storms. The consequent action of the waves may disrupt the bottom sediment and have a major effect on the flux of chemicals. Monthly sampling misses these storm effects, which occur within hours rather than days. A fuller discussion of the greenhouse effect is provided in Chapter 9.

MINING

Deposits of metals are to be found in such quantities on the sea-bed as to make the mining of these an economic possibilty. Manganese nodules, on average the size and shape of a large potato, are to be found in concentrations throughout the world. Their recovery from the bottom is relatively easy using undersea mining techniques of hydraulic grabs, suction or continuous bucket systems. About half the nodule is manganese and the rest a variety of other metals. Nodules are not the only source of metals from the ocean bottom. The first discovery of metal-rich muds in abundance was made in the Red Sea. Here they have accumulated in 'deeps' which are depressions in the bed where water temperatures may exceed 55°C due to local volcanic activity. Precious metals, including gold, are to be found in some coastal sediments especially offshore of Australia, New Zealand and Alaska. Marine polymetallic sulphides (MPS) are mineral-rich solid deposits firmly attached to the ocean bed in undersea volcanic places. They contain up to twenty minerals in sufficient quantities to start a mining rush if they were discovered on land.

Deposits of cobalt-rich ores occur in clusters attached to bed rock rather like the crusts on bread loaves. Extraction of these deposits, economically attractive or not, involves severe agitation of the ocean bed. Disturbed sediment will reduce the amount of light reaching the bottom and, as a result, alter the biological response. The muddier waters will decrease visibility and prevent predators seeing their prey and, conversely, the prey seeing the danger of predation. As with storms the mining operation may well bring back into active life toxins which have settled in the sediments, possibly those arising from anthropogenic sources – the estimated 1 million tonnes of DDT (out of the 1½ million tonnes produced before 1970) to be found in ocean sediments are, as it were, an explosive sea mine waiting to be disturbed into violence against the marine environment. Undersea mining is still in its infancy, but one thing is certain: adequate research into the conditions on the sea-bed and the possible consequences of mineral exploitation needs to be carried out before work begins. Even so the mineral resources of the oceans should be sufficient for all if they are managed efficiently and sustainably, and provided distribution is equitable.

Extraction rights become a problem except in those areas clearly covered by maritime law. In non-territorial waters are the minerals the property of any nation? Are they, like fisheries, to be exploited on a free-for-all basis or, in order to avoid problems similar to those of overfishing, should they be controlled by international agreement? Already an attempt to control mineral extraction in Antarctica, a somewhat similar situation, has resulted in countries such as the UK, France, Germany and the USA unilaterally reserving the right to exploit minerals against the wishes of other countries. Will developing nations be denied access to mineral wealth off their shores, if such exploitation causes further degradation of the ocean environment?

THE OCEANS AS A SOURCE OF ELECTRIC POWER

It is hard to believe that exploitation of the oceans in the ways described has not been matched by a similar exploitation for electricity generation. The need for power sources alternative to the combustion of fossil fuels is apparent, yet the tremendous resource of the ocean remains almost untapped. This is not the place to go into a lengthy discussion of the several ways of using ocean power but it would be presenting an unbalanced picture of the marine environment not to make some mention of the situation.

The power potential of the oceans depends on the energy available from the up and down movements of the tides and the waves, from the forward movement of currents and from the thermal and saline gradients which exist between the surface and the ocean deep.

Tidal power involves trapping the high water behind a barrage to release it at low water to turn the turbines in a similar way to hydroelectric schemes. On a smaller scale, tidal races in restricted sea areas such as the Pentland Firth off north Scotland might be harnessed by suspending large fan turbines in the flow. This is also a possibility elsewhere in situations of fast flowing ocean currents; parts of the Gulf Stream Drift of the North Atlantic might be suitable.

Wave power can be exploited with oscillating water columns (OWC) or by using waves to fill a reservoir from which high water is released via a turbine to generate electricity. Perhaps the most famous use of wave power is that of Salter's duck, which together with other designs of rafts and clams makes use of the motion of the waves.

The difference in temperature between the warmer water at the surface and the colder water at depth can be exploited by various schemes under the generic title of ocean thermal energy conversion (OTEC). It is a potential best suited to tropical areas where the thermal gradient can be as much as 20°C. The saline gradient can also be used to cause a movement between water of different saline strength based on the scientific principle of osmosis.

The potential for the use of ocean power is there; the main difficulty is not production but distribution. To link open sea production to the national distribution grids is expensive and fraught with difficulty. For the schemes well offshore the possibility of use of the electricity on site has to be considered – factories at sea may be a possible future scenario.

LEGISLATION FOR MARINE AFFAIRS

Marine legislation is complex and difficult to enforce, that applying to territorial rights especially. Until quite recent times the oceans of the world were taken to be a vast, boundless area of the world's surface, free to all (the traditional 'freedom of the seas'), apart from the 3-mile territorial waters of each coastal country, the traditional maximum range of a cannon ball fired from the shore. In more recent times the distance has extended to 12 miles and occasionally to 200 miles.

The third United Nations Conference on the Law of the Sea (UNCLOS), 1973–82, recognized the deep sea-bed as a 'common heritage'. All 150 participant nations agreed to this concept but only 134 signed the final UNCLOS III document – not the UK, USA or West Germany. By 1990 only 44 countries had so far ratified the agreement, and 35 had rejected it. Nevertheless it did decide that 12 nautical miles from the shoreline (point to point) was territorial sea with complete national control; up to 24 nautical miles was to be the contiguous zone with limited national control which includes control over foreign vessels to prevent or punish infringement of its customs and fiscal, immigration or sanitary regulations within its territory; out to 200 miles was to be the Extended Economic Zone where national rights to economic activity (fishing, mining), scientific research and conservation could be exercised. Beyond this was the continental shelf where exploration and exploitation by the coastal country were permitted. (Figure 6.2.) This is still a recommendation and is not yet part of international law – the UK has not acceded to it or to other recommendations.

The Convention for the Prevention of Pollution from Ships (MARPOL 1983) set rules for discharge from ships in relation to distance from the shore: up to 3 miles – nil; 3 to 4 miles – treated rubbish; 4 to 12 miles – treated sewage; 12 to 50 miles – untreated rubbish and sewage; over

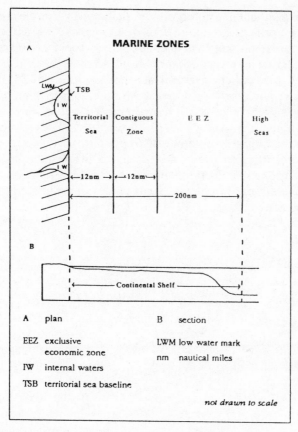

Figure 6.2 The marine zones recommended by UNCLOS III
Source: Marine Forum for Environmental Issues 1990.

50 miles – oil discharge. Ships over 400 tonnes must have tanks for waste oil.

It has taken two decades to have global recognition of the plight of the rain forests. A similar delay over concern for the oceans cannot be permitted.

It is unfortunate that the toxic effects of pollutants are not always recognized until after they have been introduced into the marine (or atmospheric) environment. It has long been held by many environmentalists that the *precautionary principle* should be applied to all new chemical compounds on request for permission to dump in the seas. This was advocated by the Second North Sea Conference on Dumping, where it was stated that 'it has to be shown to the competent international organization that the materials [to be dumped] pose no risk to the marine environment'. 'No risk' is a subjective phrase which accounts for the differences which arise between the views of the UK and other European countries, e.g. about the safety of the dumping of sewage sludge. The precautionary principle holds that those who wish to dump must show that the material presents no risks to the environment, otherwise permission for disposal at sea shall be refused. In cases of doubt the benefit must be given to the environment and not to the polluter.

What better agreement could there be for the viable future of Planet Ocean than always to consider its welfare first before the next step in its exploitation takes place? As Cousteau said: 'The future needs of society will be well served, however, only if we change our short-term mentality and often arrogant indifference to the results of our actions and focus on long-term considerations' (Cousteau 1981: xix).

The oceans are the life blood of our planet. We must not allow their vitality to ebb away.

NOTE

1 H. J. Heinz announced in 1990 that it would refuse to buy tuna 'caught in association with dolphins'. All tins of tuna sold by StarKist, a subsidiary of Heinz, would carry a 'dolphin safe' label (*Daily Telegraph*, 13 April 1990).

7 Destruction of the rain forests: Principles or practices?

Joy A. Palmer

WHAT IS BEING DESTROYED?

Forests of intrinsic value and worth. Forests of beauty, of natural harmony, of extraordinary biological diversity. Forests wherein some human beings, adapted to the rigours of high humidity and heat, have shown that they are able to dwell, not in competition with their surroundings but in a state of interdependence with them, causing no irreversible ecosystem destruction. Yet forests whose very existence is now threatened . . .

Such threat is almost impossible to believe – deriving from ignorance or greed? From malpractices or misguided principles? It seemed appropriate to begin this glimpse into the issues surrounding the destruction of the world's tropical rain forests with the words 'intrinsic value and worth'. Our minds are regularly bombarded with media coverage of the desperate plight of our planet's rain forests, with photographs, documentaries, campaigns, pleas for action and horrifying accounts of destruction statistics and their resultant threats to the world. Literature on the subject has increased at a dramatic rate in recent years and almost always sets the stage for discussion with an overview of sound yet utilitarian reasons for global concern. These will inevitably include the maintenance of biological diversity, climatic stability, protection of food and medical supplies, conserving other forest products and resources, protection of soil and water and the survival of indigenous people. Rarely are ethical perspectives highlighted amongst the myriad of complex suggested solutions to, or issues surrounding, the disappearance of the rain forests.

Perhaps this account will be as guilty as any, for inevitably the present-day utilitarian complexities of the topic must be addressed. As with most other ecological disasters threatening the future of our planet, the critical problem underpinning the forest crisis is a conflict of interests. Such conflict derives only from the dawning of that time when the emphasis of the activities of *Homo sapiens* became intrusive by nature. The form of such intrusive practices must indeed be understood. Yet beyond such practices lie values and principles which transcend what is actually a very short-lived role which present generations play in ecosystem dynamics. The rain forests of the world have survived in a relatively undisturbed state for millions of years. Since the advent of human life, generations of indigenous forest dwellers have survived by living in total harmony and supportive interaction with their environment. The world's most important natural resource, namely its diversity of biological species, has evolved in response to the undisturbed forest ecosystem. A glimpse into the unspoilt forest environment will reveal a tableau of colour and extraordinary beauty, of fascination and restless harmony, as suggested in the first documented description of a rain forest, attributed to Christopher Columbus, who claimed in 1492 that he 'never beheld so fair a thing; trees beautiful and green, and different from ours, with flowers and fruits each according to their kind and many birds'.

Yet it is the same place which has been called the 'nightmare jungle', the site of indescribable discomfort to Caucasian races, of leeches, mildew, mosquitoes: a hot, humid atmosphere which inhibits the tolerance of those adapted to western culture and which provided the setting for the intense suffering of prisoners in the Asian theatre of war in the 1940s and the Vietnam holocaust of the 1960s.

Today we see a very different picture from Columbus's tranquil description: one of threat, ugly devastation and destruction, largely brought about through the fairly recent activities of human beings. Such activities have paid little regard to such fundamental principles as concern for others and for the environment on which we depend – substituting, instead, practices of a destructive kind.

The term 'tropical forest' embraces a number of the world's forest areas. These include *evergreen forests*, also known as rain forests, which receive a large amount of rainfall – some 400 cm each year – with trees of such density that their leaves and branches form a closed canopy allowing very little light to reach the ground; *moist forests*, also having a closed canopy, but with slightly less rain than the evergreen forests; *deciduous forests*, having four to six months without rain and trees which are deciduous during this period of time; and finally *open dry woodlands*, which receive litle rain – probably less than 100 cm each year. It is the evergreen and moist forests which are generally known as tropical moist forests, and which are the subject of current and unprecedented ecological concern as they rapildy convert from vast expanses of living forest into barren wastelands.

The major locations of moist or wet forests are in South America, Africa and the islands of south-east Asia, on those areas of land which lie between the tropics of Cancer and Capricorn. The fact that there is little continental land in this equatorial zone is itself a limiting factor in rain forest distribution. Less extensive forest zones occur in Australasia, the Far East and the Indian subcontinent. Three countries alone – Brazil,

Indonesia and Zaire – share almost 50 per cent of total world closed-canopy tropical forest land. Numerous books document the horrific and often conflicting statistics of forest destruction rates; a conservative estimate is that some 900 million hectares remain out of an original 1,500 million (World Commission on Environment and Development [WCED] 1987). The rate of loss is estimated at around 11 million hectares every year. If the world has 'allowed' half of this wealth to disappear, complex and fundamental ethical questions must inevitably be raised. Is the remaining half even more valuable? Can this continue to be eroded simply because its value is not given significant enough status? Is its erosion due to ignorance, greed or inevitability? What are the principles and practices of solution? Ultimately, all life on our Earth is dependent on certain basic elements which include the tropical forests. If such questions are not considered and solutions successfully implemented, the remaining forest zones will have disappeared in only four decades from now, with devastating consequences.

> Our generation will have presided over the greatest extinction of living things since the ecological catastrophe that wiped out the dinosaurs. . . . [T]his biological holocaust is burning great holes in the web of life that sustains us all. In the short term, it threatens the lives of over a billion people, as their water resources dry up and their land turns to dust. In the longer term, it will change the world's climate – and the history of our planet.
>
> (Earthlife 1986)

Clearly, one of the key issues associated with rain forest destruction and one of unprecedented urgency is the threat to biological diversity. Conservation of species diversity is vital to human survival, in part because wild plants and animals supply foods, medicines and essential raw materials for industry. They are vital from a genetic perspective, for future improvement of crops and livestock and for the continued research into and development of new medicines and products. Animals and plants also provide invaluable services such as pest control, the natural decomposition and recycling of waste materials and soil preservation/flood control mechanisms. Tropical regions contain more than two-thirds of the world's species (up to 30 million) (Raven 1988), and threats to the forests place a sentence of potential extinction on all forms of life. One hectare of forest may have between 100 and 200 tree species. It is possible that only a single tree from a particular species will be found on a hectare of land. Other varieties of plants are present in great quantities – 7,000 types of flowering plants have been identified in west African forests, 40,000 in Brazil and 13,000 on the island of Madagascar. The vegetation feeds and provides shelter for huge numbers of invertebrates and larger animals. As with plants, there is wonder and fascination in the quantity and variety of animal species rather than the great numbers of any one type. A small patch of forest 6 km square may house some 40,000 species of insect. Amphibians and reptiles demonstrate curious adaptations (media photographs and films capture the 'parachuting' frog and the 'flying' snake). The same 6 square km may sustain 400 varieties of bird – birds of paradise, toucans, eagles and hummingbirds – and some of the world's most fascinating, colourful and rare mammals: the well-documented gorillas, tamarins, jaguars and tapirs, all fulfilling their essential role in maintaining the total life and natural balance of the ecosystem.

As tree destruction continues many species will be lost outright. Countless thousands will be reduced to populations on the brink of extinction, with the attendant problems of loss of genetic diversity and increased vulnerability to the many factors that threaten long-term survival. As the report of the Brundtland Commission states:

> the diversity of species is necessary for the normal functioning of ecosystems and the biosphere as a whole. The genetic material in wild species contributes billions of dollars yearly to the world economy in the form of improved crop species, new drugs and medicines, and raw materials for industry. But utility aside, there are also moral, ethical, cultural, aesthetic, and purely scientific reasons for conserving wild beings.
>
> (WCED 1987)

Paradoxically, it is the very biological diversity and complexity that make rain forests unique which also make them very difficult areas for scientific research. It is a well-established fact, however, that many species are rare and in danger of extinction. All species seem particularly vulnerable to habitat alteration – many have high levels of habitat specialization, and because of local distribution patterns, many species will be lost if only a minute patch or single layer of forest is destroyed. The whole is a most delicately balanced system. Any interference will have a catastrophic 'domino' effect upon the interdependency of species and specialization of habitats and relationships. It is essential that the *total* structure of the forest is protected. Whilst there is a scarcity of scientific research programmes (it is estimated [National Research Council 1980] that there are no more than 1,500 professional scientists in the world competent to classify the species found in humid tropical forests, and their number may be falling because of declining funding), perhaps far more attention should be turned to the cultural backcloth to biological diversity and to the deeply ingrained knowledge of those indigenous people who have lived successfully in the forests for generations. Only native cultures can demonstrate traditions of forest regeneration and the exploitation of the ecosystem for food, medicines and other essential products without long-term disruption or destruction. As argued by Myers (1984), the tribes of the forest are perhaps the key to understanding tropical biological diversity, and this 'human key' to plants and animals is an integral part of the ecosystems we seek to conserve. Unfortunately, forest people face a survival crisis as imminent as that of other forms of life. Their cultures and traditions, developed over generations, may disappear within decades, taking with them irreplaceable knowledge and understanding.

The climate of an undisturbed rain forest is predictable. There is little variation in temperature from month to month, and also little variation between the average highest temperatures and average lowest ones. This is particularly so below the forest canopy where temperatures are more constant than in any other ecosystem on our planet. Once again, this affects all animals and plant life. Species are adapted to a constant temperature and any significant change such as partial canopy removal will have far-reaching results, including raised daytime temperatures, cooler nights and loss of moisture through evaporation. Inevitably there will be an adaptation problem for life forms. Whilst rain forests are always wet, there is a much greater variation in daily and monthly patterns of rainfall than there is in temperature. In most areas, the majority of the rainfall occurs in a small number of heavy storms. Often there are significant variations from month to month. Some months may have well over 300 mm of rain, others only 180 mm. Probably the word 'only' is misleading – such ecosystems are remarkably 'thirsty' places, needing at least 100 mm of rain a month in order to thrive. If the level falls below this, trees and other plants will inevitably use up reserves of water from within the ground. Most rainfall occurs during the afternoon and early evening, raising humidity levels significantly. Quite a distinct daily pattern takes place, whereby the early-morning sun heats up the forest and the ground surface. As the rising warm air cools and condenses, torrential rain storms occur, dwindling into a dry night-time when clouds disperse and heat is lost into the atmosphere. Once again, there is a crucial link between the problems of rain forest destruction and climate. The torrential downpours can be 'caught' and absorbed by the mass of dense vegetation. If any portion is removed, the results are dramatic: rain hits the ground with tremendous force; soil is washed away, removing vital supplies of nutrients for the plants; and there is a great likelihood of flooding. The soil that remains is of little use for growing crops; what has been washed away causes rivers to be blocked with silt, thus contributing to extensive flooding. Forests contribute significantly therefore to agriculture and climate patterns. When rainfall is regular, people and agriculture may depend on it. Destruction of forests means loss of irrigation and interference with rainfall cycles. Together with the build-up of carbon dioxide in the Earth's atmosphere which results from forest removal, this problem also contributes to the 'greenhouse effect' or general warming of the global atmosphere, a phenomenon with potential for devastating effects on our planet as documented by Neal in Chapter 9.

Rain forest soils are actually very poor at storing the nutrients which plants need for food. Organic nutrients arising from the process of decomposition are removed in enormous quantities because there are so many plants to absorb them. Plant nutrients move through the decomposition cycle far more rapidly in the rain forest environment than they do in other ecosystems. The net result of this is that very few nutrients are stored in the soil. There is also a very poor store of mineral nutrients from underlying solid geology; minerals are soluble in water and so the rain passing through the soil washes them away in the leaching process.

It seems paradoxical that the soils of the forests are so poor or infertile, when they actually support such a wealth and diversity of plant life. As most of the nutrients are stored within this vegetation rather than the soil, the effects of tree removal are obvious, and their consequences gravely serious.

Reference has already been made to the concept of adaptation. Two further features essential to an understanding of the total forest ecosystem are renewal and fragility. Forests naturally renew themselves through a process of succession involving various stages of natural plant growth. Any scheme aimed at conservation of forests should follow this natural pattern. When gaps appear in the canopy, it is important that sufficient seeds of the correct type are available at each of the three key stages. If the aim is to regrow a forest in a cleared area, then this natural succession can be imitated by planting the appropriate seeds at each stage. The notion of fragility may seem a strange concept to associate with a huge and mighty forest – in this context, it refers to the fact that rain forests are incompetent at coping with or recovering from disturbance. They are in fact one of the most fragile ecosystems on our planet. Fragility is inevitably related to adaptation – species which have adapted to special, complex and interrelated conditions are vulnerable to change. Furthermore, because of this characteristic of interdependence and the unity of the forest, if there is a change in any element it will inevitably have a far-reaching effect on many other aspects of the ecosystem. Interdependence plus sensitivity means fragility of the unit as a whole. Any conservation attempts must take account of the entire ecosystem and the environmental and ecological conditions which surround it.

Predictably, what is utilitarian generally predominates in an overview of the 'value' of tropical forests. The trees themselves, their chief component, are an obvious and valuable source of revenue from timber and other forest products. Highly prized woods include teak, mahogany and rosewood. To the value of the timber may be added the advantage of low labour costs. Furthermore, the sale of timber provides immediate benefits for those in a position to supply it. For many developing nations, hardwood timbers are a major source of income and large quantities of wood are also used locally for fuel purposes.

In any list of rain forest goods and products, timber may be joined by fruits, vegetables, spices, nuts, medicines, oils, rubber and many other industrial materials. A less obvious 'product' is advancement in disease and pest control. Many common food plants originated in the tropics and their related wild species have valuable genetic materials that may be used to 'modify' crops throughout the world in order to produce new pest- and disease-resistant varieties. Barley in California, for example, and sugar cane in the south-east USA have been modified in this way with the aid of genetic information from wild tropical varieties.

Forest materials can be used to make a huge range of goods that are regularly taken for granted, ranging in diversity from golfballs, glue, nail varnish and chewing gum to toothpaste, deodorant, shampoos, mascara, lipstick and wickerwork furniture. For some, it has not been necessary to destroy a forest. For others, the price has been high.

No reference to the forests' value would be complete without re-emphasizing the human factor; over hundreds of years, indigenous societies have developed traditions, understanding and unique cultures already alluded to. Can any value be placed upon such wealth? The estimated human rain forest population today is around 200 million and comprises various tribes. For example, in the African rain forests are bushmen and pygmies; American Indian tribes inhabit the forests of South America; and in south-east Asia are found pygmies (in the Philippines), Sianh Daya people (Borneo) and the Biami and Gibusi peoples (New Guinea).

Rain forest people are essentially 'uninterrupted' cultures; that is, they have developed very distinctive ways of living which until recent times have not been subject to those dramatic changes in lifestyle which have transformed industrial nations. Their approach to life is at a subsistence level, in complete harmony with nature – they live, for example, in small and widely spaced communities so that no intrusive demands are made on food or resources from one patch of forest. Rain forest people are one of the few societies in the world who dwell on Earth without exploiting it to the point of destruction. Whilst the 'great' industrial nations have made significant advances in science and technology, and have 'mastered' the planet and its resources at the forefront of their reality, so forest dwellers such as the Yanomami Indians have deepened their harmony with and understanding of the Earth's natural equilibrium in order to survive, sustaining a position as an interrelated part of the total environment.

WHY . . . AND SO WHAT?

The causes of deforestation are multiple and complex. The chief direct causes are agriculture, logging for timber and industry, cattle ranching and large-scale development projects. Interlinked with these are a range of more indirect causes, including human population growth, ever-increasing demands throughout the world for forest products, unequal distribution of territory in the tropical lands and inevitably the economic situation – the developing world has a great debt to pay. Underpinning many of these causes (and space inevitably does not allow for a detailed commentary on each) are the immediate needs of those who live in rain forest nations. Many acres of forest are felled because this action provides instant benefits; the intrinsic worth of the forests is perhaps no price to pay in people's minds for what is gained when they are removed – space for valuable farmland and desperately needed cash income from timber sales.

Another useful model for analysis of the complex causes of deforestation is to understand those which come from within the forests and those generated from the outside. Within the forests, human population growth is a major factor; given that the natural ecosystem is not otherwise disturbed, one village may double its population within the next twenty-five years, especially if hygiene and health care continue their present trend of improvement. This trend is often accompanied by socio-economically forced migration, as the dispossessed try to secure a livelihood on the forest margins, for example in Brazil and Indonesia. In order to accommodate these population increases and the resultant space crisis, settlement boundaries will inevitably encroach deeper and deeper into the forest. On a worldwide scale, population explosion will cause vast tracts of virgin forest to be cleared or disturbed. Added to this will be direct threats from outside the area – notably the removal of timber and agricultural and industrial products. Each year, some 5 million hectares of tropical rain forests are cut for timber and related purposes, including paper pulp. Losses for this purpose are greatest in west Africa and Asia, and are gaining in momentum in the Amazon area of South America.

The economics of the issue are inevitably complex and related to the colossal debts of developing countries to industrial nations. Many nations borrowed large sums on the late 1970s in efforts to offset rising oil prices and to protect their economies. The repayment of loans was rendered impossible as a result of the gloomy combination of global recession and rising interest rates. Brazil, for example, owes some $100 billion, and around 40 per cent of its export earnings are channelled into debt repayments. Many such debts are now being written off by creditor banks. There seems irony in the fact that many of the loans resulting in debts were for large-scale development projects such as hydroelectric dams, roads and industrial sites – in themselves a direct threat to the rain forests. Brazil, for example, lost 25 per cent of its Amazonian forests to development initiatives in a ten-year period. The issue of donation or withholding of cash for development is complex and highly controversial. Much is documented, for example, about the controversial role of the

World Bank, which has financed many large-scale tropical development initiatives (Shiva 1987). In September 1990, the World Bank itself published a first environmental report which acknowledges many problematic issues in past policy.

Trees mean wealth: instant cash benefit from resources in world demand. Japan, the leading importer of tropical hardwood timber, receives 40 per cent of the world's total exports, followed in second place by the USA. Apart from the sales of hardwood for the manufacturing industry, timber is channelled into industrial fuel requirements and – even more controversially – into paper manufacture. Consider the ethics of a new 'progressive' technique enabling industry to convert hardwoods into paper pulp; one Japanese example concerns a company which manufactures paper packaging from imported rain forest timber of Papua New Guinea.

Inevitably in this technological age, major advances have been made in logging techniques since the days when a group of men may have spent an entire day felling one tree with axes. Today, whole trees are reduced to wood chips within seconds – progress? Apart from the impact of direct deforestation, indirect problems are inevitable: perhaps only two or three trees in a certain location are suitable for logging; the passage of modern equipment, the crashing down of felled specimens, the dragging away of logs will leave a trail of colossal devastation; neighbouring trees will be damaged; and the impact on bird and animal life will be incalculable. It seems quite extraordinary that, despite the damage, selective logging for hardwoods is often not counted in statistics of tropical deforestation. (Figures produced by the UN Food and Agriculture Organization in 1982 exclude logged areas.)

One of the most significant causes of deforestation is agriculture. Traditional shifting cultivation or 'slash and burn' farming is responsible for the clearing of some 8 million hectares of land including primary forest. This long-established system involves the clearing of forest plots by slashing vegetation which is then burned, allowing bonfire ashes to add fertility to the poor-quality soil. Crops will then be grown for two to three years until the soil nutrients are depleted, when the farming family moves on to a new forest patch, allowing the original a lengthy fallow period for recuperation and secondary growth. In the context of traditionally long rotation periods, this land-use system is self-sustaining. In the present-day climate of dramatically increasing population density, the system fails; fallow periods are insufficient for the regeneration of soils and nutrients, soil erosion is widespread, and an agricultural wasteland ensues. It is estimated that current misuse of land in this way is responsible for 50 per cent of tropical deforestation in Asia and as much as 70 per cent in Africa. This is a perfect example of how 'progress' has destroyed a traditional system that was in harmony with a sustainable ecosystem.

On a much larger scale, land clearing for cattle production causes very extensive forest destruction, particularly in Latin America, where it is claimed 'forests become burgers'. In the past twenty-five years, more than 25 per cent of this region's forest has been turned over to grassland. Its grazing cattle supply the US burger industry, with understandable economic incentives; Central American beef comes at considerably lower cost than home-produced – forest grazing land is 'free'. Once grazed, the land is productive for only a short period of time. Erosion and nutrient depletion are serious consequences, exacerbated by continual pressure from cattle hoofs and vehicles. As useless land is abandoned, more forest is cleared, and so the vicious circle continues.

Whilst single causes of deforestation may be identified and analysed, local interactions are inevitably far more complex than a simplistic over-view suggests. The process as a whole is a derivation of the entire pattern of world development since the colonial era, and most tropical countries have economies that demonstrate a number of parallels that contribute to a greater or lesser extent to the deforestation process. As the causes are many-faceted and complex, so too are the consequences. The key question to be borne in mind is, to what extent is it understandable (even excusable?) that rain forests are destroyed to stimulate economic growth and expansion in developing lands?

A summary of the well-documented consequences (as ever, utilitarian in emphasis) would probably include five major areas of concern: namely, biological diversity and species loss; impact upon indigenous cultures; effects on soils and global climate; loss of products; and social and economic considerations. Many of these have already been alluded to. Let it suffice to stress that they are a *total set* of interrelated effects – the biological holocaust; the displacement of those who practise coexist-ence with nature rather than its destruction; loss of cultural traditions, knowledge of forest ecology and quality of life; environmental deterio-ration; poverty, conflicts and political instability; global warming, with ulimate life destruction. Few doubt that the consequences are incalculable at best, and they signal the end of Planet Earth at worst.

Perhaps in an ideal world, forests would be preserved with immediate effect and put to sustainable economic use. Just how realistic sustainable development models prove to be in the long term is largely a matter of conjecture, but it can only be encouraging that they are emerging – with impact upon the very delicate balance that exists between short- and long-term gain. To achieve some degree of realism, the major trend of such models needs to be a dramatic slowing down of deforestation rates. This is already being achieved in the Brazilian rain forest, where destruc-tion declined by 30 per cent between 1989 and 1990 following a campaign against tree burning.

One of the key measures recommended by the Brundtland Report is to establish large forest reserves to protect land and species, backed up by an international agreement providing for the safeguarding of biological diversity. Other efforts to slow or reverse deforestation include the improvement of forest management, preventing large-scale developments that use resources in an unsustainable way, restoring deforested land, slowing population growth, improving agricultural practices and slowing down the demand for forest products. At the present time, less than 5 per cent of the remaining forests are protected as parks or reserves. Brazil has an established system of conservation areas that cover almost 15 million hectares; and in Costa Rica a combination of wildlife refuges, parks and Indian reserves protects 80 per cent of the remaining forest lands. On paper at least, such attempts sound impressive, but in practice they are not without problems. Too few rangers exist for efficient security and management. Also, whilst originally established in remote locations, the construction of roads and airstrips has enabled the reserves to become a focus of tourist attention. Travellers, researchers, poachers and illegal settlers all take an increasing toll.

Probably the most successful reserves are those which have been designed to benefit people as well as natural resources, having distinct zones. In the centre is a core of strictly protected land (no dwelling, agriculture, logging, mining, etc.). Around this is a 'buffer zone' for low-density inhabitation by local people, and activities such as research, tourism and education; and beyond is a 'transition' zone in which agriculture and forestry take place in a sustainable manner, providing food and income for reserve dwellers.

Management, restoration and progress towards a sustainable future can obviously take place at a variety of levels. Governments, international agencies, large non-governmental organizations and individuals all have a role to play. There is increasing evidence of concern and action at global level, particularly in the post-Brundtland era, although this is not without scepticism and criticism. One of the plans receiving early support from the Brundtland Commission is the controversial worldwide Tropical Forestry Action Plan (TFAP), devised and developed by the World Resources Institute together with the United Nations Development Programme and the World Bank. This focuses on action areas including the preservation of tropical forest ecosystems, reforestation, commercial forestry and the production of fuelwood.

TFAP initially received widespread support from both governmental and non-governmental organizations (including over sixty industrialized nations). Supporting organizations in developing countries pin hopes on the view that such powerful institutions as the World Bank and the UN ought to be successful in effecting change towards a sustainable future. Others articulate total rejection of the plan's potential in the firmly held belief that exclusive concern with commercial criteria has been the main cause of ecological destruction and cannot therefore reverse the trend. A full debate on TFAP is provided in Hurst (1990) and criticisms are also debated in the *Ecologist* (1990, 1991). Without doubt, many reject it as it stands, and in 1987 an alternative tropical forest action plan was outlined (Hurst 1990) by the World Rainforest Movement, a coalition of Third World and western environmental groups. This alternative plan identifies four interrelated components that require change: ending the debt crisis; reforming trade patterns; halting environmentally destructive development projects; and landownership reform. Its aim is to present a framework for discussion.

Whilst the World Resources Institute adopts the belief that it is the people themselves – the rural poor – who are key agents of forest destruction (for agriculture, fuel, etc.), opponents of TFAP are of the firm conviction that it is market economics which erode forests rather than their inhabitants.

An interesting commentary on the two views of the forest crisis is provided by Shiva (1987). Criticism of this kind forms part of a substantial and articulate opposition to the Brundtland Report's view of development. A number of developing country and environmental organizations and individuals claim that the commission's definition of development is highly contentious. They argue that analysis and recommendations are based on its particular conception of development and, furthermore, on economic growth and a prescription for imposing a western standard of living on all of the world's people, irrespective of their needs and desires.

It may seem illogical to doubt the essential nature of worldwide initiatives on development and the importance of major projects, particularly those that are agreed and supported by all parties concerned. Nevertheless, deep reservations about the wisdom and success of these initiatives exist and cannot be ignored. I wish to articulate a twofold doubt. First of all, perhaps this seemingly welcome shift to the macro level of solution may in fact be counter-productive, or even doomed to reinforce competition. The fact that the world is becoming increasingly united in its efforts to save the planet is cause for celebration. Sustainable development initiatives must be encouraged, but it is arguable that these are best managed within local communities, with an appropriate shift in resources to that end. Small community motivations may best meet needs of rain forest areas without compromise or exploitation. An individual prescription may well be better medicine than a global remedy.

Second, it concerns me that the explicit and dominant thrust of the development debate remains with such issues as relative weights of problem impacts and causes, competing interests, resource management and resolution of conflicts between human life and that very system upon which it depends. At the root of this debate lie fundamentally differing viewpoints on the key questions raised in this chapter: what is being destroyed, and why? Even in this scientific age we may have but a faint picture of what the rain forests truly are, and of their real value and worth. Indeed, their destruction is probably easier to understand than its subject.

Attempts to tease out that driving force which motivates and fuels the destructive process require a consideration of the powerful forces of ignorance and stupidity versus exploitation and elitism.

If interests are conflicting, both between nations and between *Homo sapiens* and the environment, promoting a spirit of competition will provide no solutions. The implementation of an essential 'ecological approach' needs to emphasize the futility of conflict at all levels. The scale of damage caused by past action seems to have escaped the good sense of logic and reason. Hopefully, the way forward will take account not just of informed and sustainable *practices*, but also of those *principles* of co-operation, respect, far-sightedness, tolerance and trust which demonstrate genuine concern both for the tropical nations of today and also for future generations. Furthermore, there is the hope that it may encapsulate a genuine concern for preserving the intrinsic value of our Earth's forests.

8 Environment and water resources in the arid zone

Robert Allison

Think of arid lands and the mind focuses on the hot, dry desert wastes of the world (Figure 8.1). Dry countries are increasingly demanding worldwide attention, and, as Goudie (1990) notes, one of the biggest environmental issues of the last two decades has been the question of desertification. Dry areas subdivide into extremely or hyper-arid, arid and semi-arid zones. In the former, water is all but absent and the land generally devoid of human habitation. Numerous schemes have been proposed for quantifying the threefold division. Heathcote (1987) suggests that hyper-arid areas receive less than 25 mm of rainfall per year, arid areas between 25 mm/yr and 200 mm/yr and semi-arid regions between 200 mm/yr and 500 mm/yr. In comparison, Great Britain receives, on average, 1,000 mm of rainfall per year and tropical countries, such as parts of south-east Asia, 5,000 mm/yr.

The dry parts of the world are centred on the tropics, comprising large parts of Africa, North and South America, India, Pakistan and the Middle East (Table 8.1) More than 80 per cent of the total is located in Africa, Australia and Asia alone. At the heart of the arid zone are five major deserts: the Afro-Asian, North American, Atacama, Namib-Kalahari and Australian (Figure 8.2). It is more the desert fringe areas, zones of marginal human culivation and habitation, which are of relevance here. The ecological stability of arid and semi-arid environments is closely linked to the local hydrological flux and the balance between water surplus and water deficit. Eleven nations fall squarely within this category; their territory is all arid or semi-arid. Another 23 countries have over 75 per cent of their land affected by aridity, and a further 5 nations have 50 per cent of their land within the arid or semi-arid zone (Table 8.2). In addition, 27 states are partly affected by water deficit problems.

The human impact of desertification is most pronounced in Africa and Asia (Table 8.3). These two continents include 70 per cent of all arid areas and 82 per cent of the 281 million people for whom dryland problems are a fact of everyday life. It is also interesting to note that

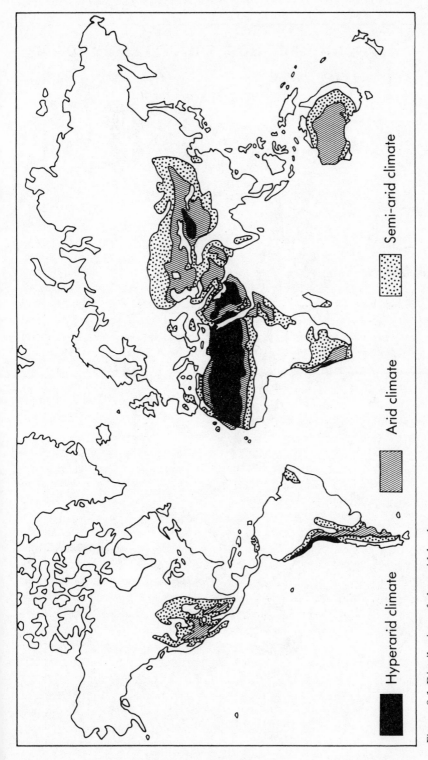

Figure 8.1 Distribution of the arid lands
Source: Drenge 1983.

Hyperarid climate

Arid climate

Semi-arid climate

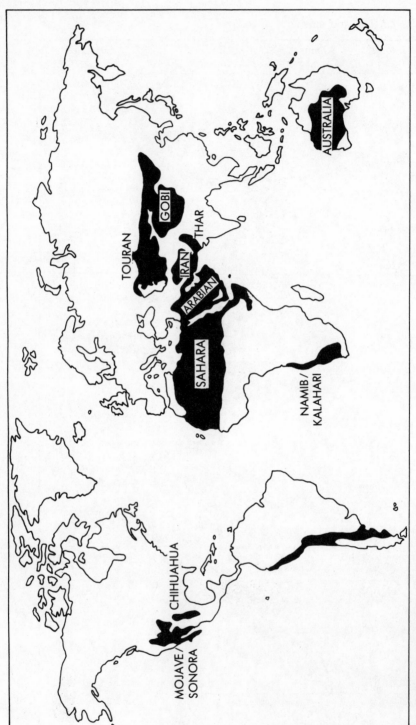

Figure 8.2 Major deserts of the world
Source: Goudie 1984.

Table 8.1 The arid nations

Description	Number	% of nation arid or semi-arid	Countries
Arid	11	100	Bahrain, Djibouti, Egypt, Kuwait, Mauritania, Oman, Qatar, Saudi Arabia, Somalia, S. Yemen, United Arab Emirates
Predominantly arid	23	75–99	Afghanistan, Algeria, Australia, Botswana, Burkina Fasao, Cape Verde, Chad, Iran, Iraq, Israel, Jordan, Kenya, Libya, Mali, Morocco, Namibia, Niger, Pakistan, Senegal, Sudan, Syria, Tunisia, Yemen
Substantially arid	5	50–74	Argentina, Ethiopia, Mongolia, South Africa, Turkey
Semi-arid	9	25–49	Angola, Bolivia, Chile, China, India, Mexico, Tanzania, Togo, USA
Peripherally arid	18	<25	Benin, Brazil, Canada, Central African Republic, Ecuador, Ghana, Lebanon, Lesotho, Madagascar, Mozambique, Nigeria, Paraguay, Peru, Sri Lanka, USSR, Venezuela, Zambia, Zimbabwe

Source: Drenge 1983.

Table 8.2 Distribution of dry lands[a]

Region	Area (10^6 km^2)	% of total
Africa	17.3	37
Asia	15.7	33
Australia	6.4	14
North America and Mexico	4.4	9
South America	3.1	7
Europe	0.2	0
Total	47.1	100

Source: Paylore and Greenwell 1980.

[a] Including hyper-arid, arid and semi-arid zones.

Table 8.3 Populations affected by desertification

Region	Affected area (10³ km²)	% of total	Affected population (000,000s)	% of total
Africa	7,409	37	108	38
Asia	7,480	37	123	44
Australia	1,123	7	0.23	0
Mediterranean Europe	296	1	16.5	6
North America	2,080	10	4.5	2
South America and Mexico	1,620	8	29	10
Total	20,008	100	281.23	100

Source: Mabbutt 1984.

some parts of the world not usually asosciated with water shortage problems, such as areas in the USA, fall squarely within the arid zone.

Water-related problems in arid lands are numerous and some of the most acute environmental concerns have a direct association with the water balance. The World Bank, for example, cites ground salinity in arid lands as one of the major environmental problems confronting mankind today. Such areas of conflict are as much a direct consequence of poor water management strategies as they are of lack of water resources. Indeed many of the key hydrological concerns are inextricably associated with man's manipulation of the natural environment in places where ecosystems are delicately balanced. Areas under potential danger cover over one-fifth of the land surface of the Earth, threatening the livelihood of over 16 per cent of the world's population. Those at risk are those most inadequately prepared to minimize the potential environmental damage. Lack of finance, little or no technologcial support, inability to perceive the threat to livelihood and woefully inadequate strategies for minimizing risk all work to compound the problem.

Despite the emphasis on aridity and lack of water, even the driest parts of the world receive some rain, albeit on a highly irregular basis. A severe lack of water prevents human habitation. The Empty Quarter in the eastern Rubal Khali, Saudi Arabia, is a case in point. In the absence of man, arid zone problems associated with the water balance are virtually absent; the natural environment functions satisfactorily within an order determined by low-frequency, low-intensity water inputs. It could be argued, as is all too frequently the case with environmental issues, that difficulties only rear their head when a human population disrupts the natural arid zone ecosystem. This most clearly manifests itself when people inhabit marginal areas where the availability of usable natural water is precariously balanced between surplus and deficit. In years of plenty, supply exceeds demand; in years of drought, demand outstrips supply. Sometimes even one year of inadequate supply can have serious environmental repercussions, with sequential drought years further compounding difficulties.

It is important to recognize, particularly in the context of arid zone water management, the differences between drought and desertification. The terms are ill defined and frequently their inappropriate use causes confusion. Drought is a short-term problem and a consequence of an immediate lack of water. In unpopulated parts of the arid zone drought years will pass without notice. When a lack of water affects populated areas and available water reserves dry up, serious concerns develop including crop failure and famine. Droughts, despite being short lived, arouse international emotion because of their association with starvation in Third World countries.

Desertification encompasses a different set of problems associated with land degradation over a long period of time. The causes may be climatic or anthropogenic (Rapp 1974). The latter include excessive cultivation, heavy grazing by animals, deforestation and salinization. The net result manifests itself in the degeneration of agriculturally productive land and sand encroachment (Mortimore 1987; Rapp 1987). Desertification is thus associated with inadequate water management in the arid zone but other contributory factors such as overgrazing (Noble and Tongway 1988), vegetation changes (Le Houérou 1977) and the ease with which wind can mobilize topsoil (Smalley 1970) play a part in compounding the degradation process. Desertification is a long-term phenomenon and more difficult to solve than short-term drought.

The recognition of drought and desertification is not new. The troubles in central Africa were realized as early as 1926, for example, when Renner identified desert encroachment from the Sahara, a reduction in annual precipitation totals and a gradual fall in groundwater and lake levels in Sudan. In short, the onset of water-related environmental problems in the central African part of the arid zone was confirmed in the early part of the century, but it has taken over fifty years and a sequence of famines to focus attention on the plight of those living in the area. A similar situation exists for other countries including India (Stebbing 1935) and Chad (Jones 1938).

As the drought/desertification issue illustrates, the effective management of water in hot, dry environments is constrained by a temporal framework. In the short term, over one year or less, environmental problems will be limited for two reasons. First, the natural environment possesses a degree of 'elasticity' which, providing the natural order quickly returns, will balance anomalies. Second, many of the most acute concerns develop in the medium and long term. Although medium- and long-term effects have to be tackled at the earliest opportunity to minimize environmental degradation, serious consequences take a number of years to evolve.

With the above points in mind, four specific areas require further attention in terms of associations between water and the environment in the arid zone. These are the nature of water inputs, runoff characteristics and associated land erosion, the use and abuse of groundwater reserves and irrigation, and the associated problem of ground salinity.

WATER INPUTS

Despite being characterized by rainfall of less than 500 mm/yr, the most significant water inputs in the arid and semi-arid zones are from precipitation. Dryland rainfall is highly variable in two ways. First, most precipitation occurs during specific parts of the year and even then it is not guaranteed. In the dry countries of central Africa, for example, 85 per cent of annual rainfall is received between July and September (Grainger 1990). Second, the total water supply can vary from year to year and it is possible to have a string of dry or wet years in succession. Rainfall is a consequence of either the passage of low-pressure weather systems or the presence of small, local convective cells. The former result in rainfall over wide areas, while in the latter precipitation is more localized. Individual storms are frequently intense and of short duration. One event can supply a volume of water exceeding the mean annual rainfall. Rainfall inputs over 1 mm/min or, during extreme events, 100 mm/hr are not uncommon. Mean annual precipitation figures are therefore relatively meaningless and parameters such as rainfall effectiveness, intensity and the rate of generation and amount of overland flow are more relevant (Table 8.4).

In the short term, in other words less than one year, precipitation is usually centred on a monsoon season. In the medium term, over one to one hundred years, wetter and drier cycles may exist. Mean totals calculated from a number of consecutive years of data will show temporal patterns. Medium-term wet and dry cycles have implications for human activities such as agriculture, particularly if mean annual rainfall amounts lie close to the boundary between water surplus and water deficit conditions. Water deficit years will result in crop failure. A string of deficit years in succession will have repercussions for human populations in terms of food availablility. In addition, overcultivation and overgrazing during dry periods may result in permanent damage to the soil and irreversible land surface degradation. Changes in water inputs over long time spans of more than 100 years are difficult to assess. Reductions in water inputs manifest themselves over many years and will be accompanied by variations in climate elsewhere in the world. Even during periods of time when water supply is reasonable in arid lands, the high temperatures will result in excess evaporation over precipitation.

The two most important uses of rainfall reaching the ground surface in the arid zone are groundwater recharge and immediate consumption for human, agricultural or industrial purposes. Plant growth in arid lands,

Table 8.4 Arid zone climatic indices

Index	Originating Scientist
Rain factor index $= P/T$	Lang
Aridity index (I) $= P/(T + 100)$	De Martonne
$I' = \dfrac{P/(T + 10) + 12p\,(t + 10)}{2}$	De Martonne
Precipitation/saturation deficit ratio $= \dfrac{P}{SD}$	Meyer
Precipitation effectiveness index $= \dfrac{P'}{E}$	Thornthwaite
Rainfall efficiency index $= \dfrac{100P/E = 12p'/e'}{2}$	Capot-Rey
Index of dryness $= \dfrac{R\,(1/yr)}{L \times P}$	Budyko

Sources: Lal 1987; Nir 1974.

Key:
P   = mean annual precipitation
P'   = precipitation
p   = driest month total rainfall
p'   = wettest month total rainfall
T   = mean annual temperature
T'   = mean monthly temperature
t   = driest month mean temperature
E   = evaporation
e'   = wettest month evaporation
SD = saturation deficit
L   = latent heat of vaporization

for example, is related to water availability. The extent to which incoming water resources from precipitation can be utilized depends on a broad spectrum of parameters such as rates of evaporation, soil conditions and the precise nature of the rainfall event. High-intensity storms will result in fine soil particles at the ground surface being washed into pore spaces, reducing infiltration and limiting groundwater recharge and water availability in the root zone for uptake by plants. The rate at which water can seep into the ground, or infiltration rate, is rapidly exceeded during high-intensity storms. Once the limit or infiltration capacity of the surface has been reached, water will begin to flow over the surface of the earth as runoff.

RUNOFF AND LAND EROSION

It is perhaps strange to be considering problems associated with runoff, or overland flow, in the arid zone due to the lack of ground surface water for much of the time. However, the intensity of individual storms frequently results in large volumes of water moving across the land surface. The environmental problems created by large volumes of surface water are considerable and include soil stripping, sediment redistribution, falling crop yields and damage to man-made structures. A complex suite of problems confronts all parts of the arid zone. Add to this the fact that in many areas effective management and conservation practices either are not used or have not been successful and the scale of the problem is further magnified.

During and after heavy rainfall flash floods will pass down dry river or wadi beds, producing a wall of water and peak flow conditions (Figure 8.3). The speed with which floods develop is a consequence of low ground infiltration rates and an absence of vegetation to retard moving water. Flood peaks involve the transfer of large surges of energy across the landscape, often with destructive consequences. Water levels in wadis are highly irregular. Maximum energy conditions may pass downstream in less than ten minutes but a reduction to normal flow conditions may take hours and the majority of the time is characterized by no water at all. When rain falls, a resulting flood peak can travel a considerable distance and even traverse areas where precipitation is absent. The environmental problems in these rain-free areas are exacerbated because there is no warning of approaching surface water. Damage to man-made structures such as bridges and embankments (Figure 8.4), loss of dwellings built on or near to wadis, large-scale inundation of the land surface and loss of life may accompany big discharge events. A combination of evaporation to the atmosphere and infiltration through the wadi bed and banks will gradually reduce the volume of water moving downstream until it disappears. It is seldom the case that water flowing in wadis will reach the sea.

Associate with moving water is accelerated erosion, transport and deposition of sediment (FAO 1965). Water will concentrate into rills and as the depth and velocity of the flow increase so will the detachment and transport of particles (Figure 8.5). Precipitation impacting and water moving across the ground surface have the capacity to entrain large quantities of soil and other material. Erosion rates are affected by numerous variables (Table 8.5). Important rainfall parameters in determining the magnitude of erosion are drop size, drop distribution and the angle and velocity of impact with the ground surface. Properties of the ground surface including material particle size and the angle of slope play an equally important role. Clay and silt-size particles are transported more easily than sand-size particles, for example. This highlights the

Figure 8.3 Flood peak passing down a wadi channel, Sultanate of Oman

Figure 8.4 Structural damage to a bridge following wadi flood conditions, Saudi Arabia

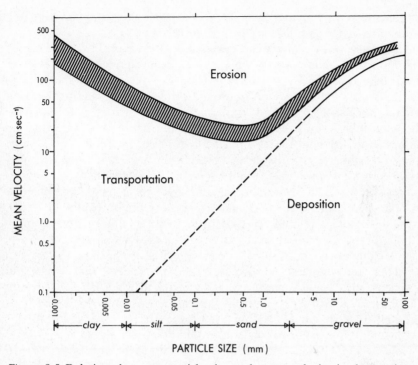

Figure 8.5 Relations between particle size and water velocity in the erosion, transport and desposition of sediment
Source: Hjulstrom 1939.

Table 8.5 Factors affecting erosion rates in the arid zone

Forces promoting detachment		Forces resisting detachment	
Rainfall	*Overland flow*	*Soil properties*	*Ground cover*
Kinetic energy	Depth	Texture	Crop residue
Momentum	Raindrop : overland flow interaction	Structure	Vegetation cover
Drop size distribution		Infiltration	
Intensity		Organic matter	
Amount		Crusting	
		Horizonation	

Source: Lal 1990.

relationship between the erosivity of water on the one hand and the erodibility of the ground surface on the other.

The nature of human activity can greatly affect erosion rates. Increasing population demands in marginal areas under agricultural production exacerbate erosion. Uncontrolled grazing by livestock rapidly destroys natural vegetation and increases ground surface sealing due to the compaction effects of large herds of animals. If overgrazing is followed by the relocation of a population, accelerated soil erosion will follow on the bare, vacated ground. Floret and Le Floch (1984) report the loss of 88.5 t/ha of soil in five months due to overgrazing in Tunisia, for example, while Alkamper et al. (1979) note irreversible damage to terrace cultivation systems in the Haraz province of the Yemen Arab Republic due to poor management practices and significant ground erosion. As much as 50 per cent of rainfall received in high-intensity storms will turn to runoff, even on carefully managed arable areas (Agassi et al. 1986).

The cumulative effect of water moving across the ground surface can be significant and large areas may be stripped clear of loose soil and sediment, increasing water turbidity values to extreme levels. Sediment concentrations in flowing water as high as 20 kg/m^{-3} in Argentina have been reported by Nir (1974), for example. Reported annual records for other semi-arid catchments include 38 $t/km^2/yr$ in Israel. In the Colorado region of the USA some catchments record sediment losses of 370 $t/km^2/yr$.

Water supercharged with sediment can modify the land surface in other ways. As the flow level drops, particularly in the lower reaches of wadis where the channel widens out over large surface areas, sediment will be deposited. Deposition is usually stratified, with the finest material being the last to be laid down. The consequent effects are numerous. Ground surface sealing will occur as an impermeable skin develops, reducing infiltration rates; crops may be buried and the elevation of wadi beds will increase. Accompanying the latter will be a reduction in wadi channel capacity. Later floods of the same magnitude will therefore come much closer to over topping the wadi banks, ultimately resulting in widespread ground inundation. In other words, a positive feedback mechanism comes into play and the problems associated with surface water flow in arid regions become self perpetuating.

In the medium and long term, the overriding concern associated with runoff, soil erosion and sediment movement is the extended period of time required for new soil to develop. The resource is non-renewable in terms of the human life time scale. In all instances therefore prevention is better than cure. Identifying potential runoff-related problems before they develop is the best way of eliminating difficulties. Preventive techniques are centred on good land use strategies. Ground highly susceptible to erosion should be left under natural vegetation. Where land use is changed for either agricultural or urban development, modifications have to balance economic returns on the one hand against the degradation of the natural environment as a resource base on the other. Good agricultural practices should include using diverse farming systems, carefully selecting crops, their rotation patterns and crop management systems and physically manipulating the soil by processes such as tillage and mulching to maintain optimum conditions for plant growth and water infiltration.

In some cases remedial action has to be taken to control, rather than prevent, erosion. Integral to this is the management of surface runoff. By reducing the velocity of overland flow erosion rates will automatically drop. Engineering structures can be built to divert flowing water away from sensitive areas and minimize gully erosion. Most importantly, land use practices can be modified and vegetation planting programmes established to reduce erosion rates. Control schemes can be highly successful providing that they are carefully defined and strictly adhered to. Once an arid ecosystem has been restored to natural conditions, prevention techniques must be implemented to eliminate the likelihood of a return to conditions of soil erosion and land degradation.

GROUNDWATER

Groundwater is the most usable water resource in the arid zone. New techniques for extracting water from below the ground have resulted in the removal of increasing volumes as well as making it possible to tap into deeper reserves. Ever better methods are available for identifying usable sources of sub-surface water. It is increasingly the case that water-bearing rocks below the surface of the Earth, called aquifers, are viable in terms of extracting the water they hold in places where the cost benefits were previously insufficient to warrant exploitation. Associated with increased groundwater extraction are numerous problems. Reduction in the level of water tables, the pollution of groundwater, ground subsidence and ground surface salinization are all cases in point.

The advent of automated groundwater extraction provides the most critical water management problem in the arid zone today. Resource exploitation takes place at rates far exceeding recharge levels. The result is the ever increasing depletion of a major dryland natural resource. Areas of the USA are currently suffering from some of the most serious groundwater problems. In some parts of the high plains of Texas, for example, the water table has dropped by over 50 m in less than fifty years, representing a 50 per cent decrease in the size of the groundwater reserve.

Groundwater problems are exacerbated in arid zone countries by a number of factors. Rising populations place increasing pressure on the available resource, in terms of both direct consumption and the amount of water required for irrigation and food production. Intensive farming methods and new techniques have been accompanied by a breakdown in traditional 'rule of thumb' water allocation principles which have evolved historically within indigenous populations. The quantity of water removed from the ground was originally based on historically derived principles which allocated fair shares to all, without depleting water resources to danger levels. The recent installation of well pumps has led to the reduction of the water-bearing potential of many sub-surface reservoirs.

Not only do water levels decrease but the extraction process results in a reduction of fluid pressure in the underground reservoir, causing compaction of the water-bearing rock and ground surface subsidence. To some extent the problem of subsidence is predictable but occasionally complicating factors introduce additional environmental hazards. Cooke and Doornkamp (1990) report an example from Texas, where heavy water withdrawal from an aquifer has resulted in ground surface subsidence of over 1.5 m at a rate exceeding 760 mm/yr between 1959 and 1964. Ground subsidence is seldom uniform through time or space and can cause damage to structures, foundation cracking and tilting to buildings. The costs in terms of damage can therefore be considerable.

Decreasing groundwater quality results from two main causes. First, as the volume of sub-surface water in an area decreases, the natural salt concentration of the water rises. In areas where the groundwater is extracted for irrigation, salts which are precipitated at the ground surface are occasionally leached away to improve soil productivity. If the water used in the leaching process passes back into the groundwater zone, water quality will drop still further. A circular, ever increasing problem thus develops. Second, if water is removed from below the ground, the encroachment of groundwater from other areas may result in a change in quality. In coastal areas, for example, seawater incursion is often significant. A saltwater wedge will advance into a region, pushing beneath the freshwater resource, displacing it and rendering the whole groundwater reserve unsuitable for exploitation.

The problems associated with hypersaline water encompass human, agricultural, urban and industrial uses. Water quality requirements for human consumption are high. A small increase in salinity or suspended sediment within the water can render the whole of the resource useless. In agriculture dissolved salts make the water unsuitable for irrigation but often the problem is not recognized until unsuitable water has been used for some time. The dissolved salts are precipitated at the ground surface, destroying soil fertility. Poor quality water is also unsuitable for animal consumption. In other words, the water resource becomes useless. As a consequence, the agricultural component of an economy can be affected. Such situations are critical in arid zone countries where food and livestock production forms a significant component of the gross national product.

In urban areas salt weathering damage to buildings can occur as a result of the capillary rise of water from the saturated groundwater zone. Water seeps into the construction material and evaporates due to high

Earth dikes

Check dams and ponds

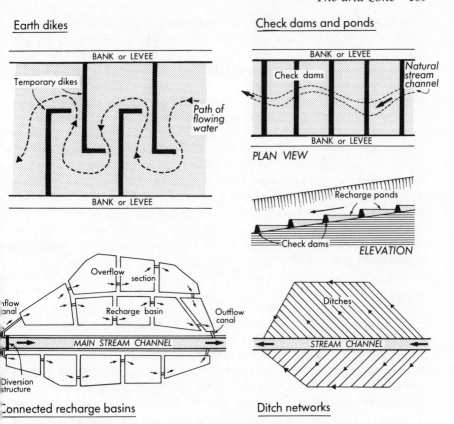

Connected recharge basins

Ditch networks

Figure 8.6 Man-made structures for impounding water, reducing runoff and promoting groundwater recharge
Source: Todd 1964.

ground surface and atmospheric temperatures. Dissolved salts in the water are precipitated out of solution in the pore spaces within the material. The growth and expansion of salts in a confined space exert increased pressures or stresses on the construction materials, eventually resulting in damage such as cracking, spalling and granular disintegration. Numerous examples can be cited of groundwater-related salt weathering problems in arid zone urban areas. The effects of salt weathering on roads have been reported in Australia (Cole and Lewis 1960) and the Middle East (Fookes and French 1977), for example, while problems associated with buildings are frequently cited for the Middle East and Gulf states (see Fookes and Collis 1975, for example).

Groundwater management practices should balance extraction and recharge rates. Returning to a state of equilibrium involves reducing the volume of water pumped out of the ground and improving recharge rates by encouraging infiltration. Man-made structures to impound water and reduce runoff (Figure 8.6) are useful in promoting groundwater recharge. By increasing the length of time water is in contact with the ground the amount percolating down to the water table rises. Alternatively water may be pumped back into the ground through wells, particularly if the aquifer is deep and the likelihood of percolating water reaching the saturated zone is small. Average well recharge rates of 2,200m³/dy and 2,700 m³/dy have been reported by Todd (1980) for California and Texas respectively. An example of water quality changes due to recharge is reported by Mandel (1977). Between 1950 and 1960 groundwater extraction in the Tel Aviv area of Israel greatly exceeded recharge, resulting in lowering of the water table and encroachment of saline water into a previously fresh aquifer. By pumping water down below the water table through a carefully designed network of recharge wells, the water degradation problem was reversed. In other words, an effective management strategy was used to alleviate a serious groundwater problem in an area reliant upon good quality sub-surface water.

IRRIGATION

In 1900, approximately 44,000,000 ha of the land surface of the Earth was under some form of artifical watering programme. By the mid–1950s the total had increased to 150,000,000 ha and by the early 1980s it had risen again to 200,000,000 ha. It is estimated that by the year 2000 about 400,000,000 ha of the Earth's surface will be irrigated. A portion of the total represents irrigation projects outside the aid zone but 81 per cent of the latest estimate comprises arid and semi-arid areas. In the drylands irrigation is not just important in terms of agricultural production but it is essential to the survival of local populations and effective national economic development. Tackling malnutrition, for example, has to be centred, in the long term, on increasing local food production, which is itself dependent on adequate water supply. The 1974 United Nations World Food Conference Report highlights the problem. The world demand for food is expected to have doubled between the mid-1970s and the year 2000, while at the same time the projected increase in cropland under cultivation is expected to rise by only 10 per cent. Thus, in the arid zone, irrigation is necessary to maintain an ecology suitable for human habitation and food production in places where rainfall alone is not sufficient to do so. Despite the theoretical advantages of irrigation, the poor management of projects frequently leads to the demise of water supply schemes within a few years of initial operation and newly cropped lands can revert to desert.

With the recent technological advances in water resource exploitation it is essential to exercise strict control of irrigation practices if a balance is to be struck between the advantages gained from watering the ground on the one hand and the disadvantages consequent upon poor resource management on the other. Water for irrigation can come from a variety of surface and sub-surface sources. Surface water is utilized in countries where seasonal differences in precipitation give rise to some months of water surplus and others of water deficit. Wallace (1980, 1981), for example, reports a project in northern Nigeria on the Kano River, where water supplied from the Tiga Dam is used to feed local holding reservoirs from which irrigation water is drawn. By controlling the redistribution of water from the holding reservoirs, crop yield and crop diversity have both been increased. In other arid areas, regions which receive a water surplus are used to supply places lacking in water. In the Sultanate of Oman, for example, convectional storms over the Oman Mountains result in green and fertile upland areas in comparison to the surrounding lowland desert. Falaj water supply canal systems are used to transport water from the mountains to the dry desert plains (Dutton 1988). Supply reliability is a key problem with all surface water. A case in point is the southern Chad irrigation project in Nigeria (Kolawole 1987). Work commenced in 1980 was designed to provide irrigation water from Lake Chad to 100,000 ha of land. With the droughts of the early 1980s and falling lake levels, no water has been available for any form of irrigation since 1984.

Groundwater forms the other source of arid zone irrigation water. Many difficulties are associated with using groundwater for irrigation including the size, extent and location of the sub-surface water resource, the level of technology necessary for getting the water out of the ground and the water quality and quality change during extraction (Singh *et al.* 1981; Wellings 1982). An additional concern is the economic viability of deep groundwater extraction and the financial implications of obtaining the water when balanced against the commercial benefits gained from its use.

The environmental impact of arid land irrigation is varied. By impounding surface water, changes will occur around and downstream from the point of water collection (Diamant 1980). The ecology of the surrounding area will change. If water impoundment schemes are large, populations will have to move and resettle. In Nigeria, for example, the construction of the Bakolon Reservoir supplied irrigation water to 30,000 ha but required the relocation of some 12,000 people, resulting in local unrest (Adams 1988). The silting up of reservoirs may be a problem if land erosion occurs upstream, affecting the efficiency and extent of irrigation. Also if silt becomes trapped in reservoirs the beneficial effects gained from floodplain inundation and sediment deposition will be lost. The Aswan Dam is a case in point. The productivity of land below the dam, along the lower Nile, has dropped significantly (Biswas 1980; Shalash 1983). Not only that, but in the Aswan case the erosion of the Nile delta has commenced; offshore fisheries have been affected; and saltwater has penetrated coastal aquifers. In other words, the consequences of dam construction are far-reaching over a wide area.

One of the biggest environmental problems associated with irrigation is ground salinization, which makes land infertile. Of all water applied to the ground surface, some infiltrates down to the root zone while one-half to two-thirds evaporates. Even with the purest of natural water resources, evaporation gradually leads to dissolved salt concentration at the ground surface. The problem is widespread (Table 8.6) and at present it is estimated that over 50 per cent of all irrigated land suffers from salt contamination (Kovda 1977). As irrigation continues the problem is compounded by increasing groundwater salinity. In other words, the volume of salts being precipitated from a given volume of water will increase, resulting in an exponential rise in problems. Amongst the most severely affected countries are Egypt (Ibrahim 1982), Pakistan (Bokhari 1980; Gazdar 1987) and Iraq (Saad 1982).

Table 8.6 Irrigated land affected by ground salinization for selected countries

Country	%	Country	%
Algeria	10–15	India	27
Egypt	30–40	Iran	<30
Senegal	10–15	Iraq	50
Sudan	<20	Israel	13
USA	5–20	Jordan	16
Colombia	20	Pakistan	<40
Peru	12	Sri Lanka	13
China	15	Syria	30–50
Cyprus	25	Australia	15–20

Source: Goudie 1990.

Solving salinity problems starts with the implementation of control measures to prevent land degradation. Control is centred on three main areas: minimizing problems during irrigation itself; reducing the salinity content of water in the ground by replacing adsorbed sodium with calcium and magnesium; and secondary measures such as the careful selection of crop types. Control during irrigation requires identifying the correct balance between water resources, the elimination of over-irrigation and identifying the correct balance between irrigation, natural drainage, return flow to the groundwater zone and the leaching processes which accompany moving water. If successful, crop water demands will be satisfied, harmless salts will be leached out of the upper parts of the soil and excess water will not build up, eliminating the accumulation of surface water, evaporation and salt deposition.

Protecting groundwater resources from increasing salinity requires a comprehensive land and water management strategy. By their nature, the processes which leach salts from the upper, agriculturally fertile soil result in the downward transfer of precipitates into the groundwater zone. Short of eliminating irrigation, management strategies are difficult to implement. Reducing deep water percolation will cut down on salts leaching back into the groundwater zone. The salinity level of return flow can be up to ten times greater than it was for the water at the time of extraction. Alternative strategies include diluting downward-percolating water or intercepting water at key points. If saline conditions cannot be fully eliminated, crops must be selected which grow satisfactorily under moderately saline conditions. Examples include sugar beet, barley and cotton grass.

In some areas salt precipitation is only identified at a late stage once crop production has been affected and land quality has declined. The only effective environmental management strategy in these circumstances is the reclamation of mismanaged land, which involves leaching with irrigation water to remove excess salts and the application of fertilizers such as calcium and magnesium to improve soil quality and structure. Drainage must also be improved to reduce salt precipitation in the upper layers of the soil and good agricultural practices have to be implemented to maintain a good ground structure and prevent the recurrence of previous problems.

It cannot be emphasized enough that the best way of controlling arid zone problems associated with irrigation is prevention rather than cure. In every instance where problems have arisen in the past, their avoidance would have been possible by implementing effective management strategies and systems. As implied by Carruthers (1981), it is not the development of new schemes that is needed to improve irrigation in the arid zone but the effective management of old ones. The operation and maintenance of irrigation schemes are frequently ignored or mismanaged once supply has been established. Consequently schemes falter, crops fail and food production ceases. In such circumstances the local population has usually come to rely on the water supply. Much mismanagement is due to supply programmes being focused on large co-ordinating bodies where lack of interaction results in inefficiency. The World Bank, for example, devotes 28 per cent of all its agricultural lending to irrigation schemes (Carruthers 1985), but due to the failure of projects in the past the Bank is keen to move away from large-scale irrigation schemes with all their attendant problems to small projects which are easily managed and have a better chance of success.

TOWARDS WATER RESOURCE CONSERVATION AND MANAGEMENT

There is little doubt that, as the next century approaches, water conservation and management will be key environmental issues in arid zone countries. Human occupancy, agricultural productivity and the economic development of dryland areas cannot be maintained without a regular and reliable supply of water. Lack of attention to water issues will result in falling crop yields and increasing desertification. The signs of a future crisis are already present in parts of the arid zone. Kenya has one of the world's fastest population growth rates. It is estimated (Biswas 1990) that over 750 mm/yr of rainfall is needed to meet the increasing demand for food but only 15 per cent of the country receives that amount of precipitation.

Not only must water be present but the policy and infrastructure for its management and use must be well defined and constructed. As the attendant problems illustrate, the inefficient use of water is ultimately as bad as, if not worse than, having no water at all. In all arid and semi-arid countries additional water can be obtained not only by exploring for new sub-surface reserves but also by improving the efficiency with which existing supplies are distributed and consumed. Many drylands suffer from excessive irrigation, for example. By controlling distribution not only is water available for use elsewhere but waterlogged ground is eliminated and salinization reduced.

Management strategies should not be viewed in isolation. Frequently differences in water supply between regions can be used to balance inequalities within a country. Water can be transferred from water surplus to water deficit areas. Those involved in central decision making are all too frequently unaware of the possibility of problem solving by integrating policies and projects. The subdivision of projects between government departments, for example, can increase bureaucratic divisions which make it impossible to develop a water resource in the most efficient manner. At their worst, projects and management plans developed centrally at the planning stage are impossible to implement due to local constraints and site-specific problems.

In some countries the tide is turning. Governments recognize the importance of water resources in underpinning a stable national economy. In the Sultanate of Oman, for example, government departments such as the Public Authority for Water Resources have been established to develop national management strategies, with regional offices opening to oversee the local implementation of policy. Such bodies are concerned with both the location of new water reserves and the more efficient use of current supplies. A good example in the latter instance is the reuse of waste water. In developing countries waste water is fast being recognized as an additional source of water. Many programmes are being developed to recycle what was previously ignored as a spent resource. Biswas and Arar (1988) go as far as to suggest that the recycling of water may have a greater impact on future water availability than any other potential solution to the water deficit problem.

In conclusion, it is clear that in all arid areas the use and abuse of water between now and the next century will be one of the key factors in determining levels of productivity, provision for human occupancy and the likelihood of desertification.

9 Air pollution – with special reference to acid rain, the greenhouse effect and ozone layer depletion

Philip Neal

It is an irony of air pollution that the cure for one problem may well exacerbate another. Much publicity, for example, has been given to the use of three-way catalytic converters in the exhaust systems of cars to 'reduce noxious emissions by 95%' in the words of a recent advertisement (Audi 1989) from a major vehicle manufacturer. It goes on to explain that as the 'engine exhaust gases pass through at around 300°C these convert nitrogen oxides, unburnt hydrocarbons and deadly carbon monoxide into nitrogen, water and carbon dioxide . . . the stuff that makes fizzy drinks fizzy'. What it does not say is that 'the stuff that makes fizzy drinks fizzy' is also the stuff which absorbs the heat being radiated back into space from the Earth, thus warming the atmosphere and so becoming the leading culprit in the phenomenon commonly known as the greenhouse effect. Unfortunately carbon dioxide (CO_2) is produced every time something is burnt, the combustion of fossil fuels included.

The moral must be that individual aspects of air pollution cannot be viewed in isolation, that the 'cure' of each problem needs to be evaluated against the cause and effect of another, and while it may be convenient to isolate individual problems their overall relationship one to the other must be borne in mind. The conclusion seems to be obvious to the comprehensive observer; clearly, as with most ills, prevention rather than cure has to be the prime objective. From government departments to manufacturing industries the onus on research and development departments is to make a full assessment of the polluting processes and environmental hazards of technological advance, as well as of the economics of the situation.

MAJOR PROBLEMS OF AIR POLLUTION

The air is a complex chemical mixture and the processes which occur within it may take place out of sight and reach of scientists. The atmosphere undergoes constant change which laboratory experiments and computer modelling are mostly unable to replicate. It may well be that the fundamentals of the atmopsheric cycle are understood, but the introduction into the air of laboratory-created chemicals, in addition to the products of incineration, increases the difficulty of interpretation. Insufficient environmental evaluation is carried out into the effects of new chemicals which soon become part of our daily lives. Chlorofluorocarbons (CFCs) and polychlorinated biphenyls (PCBs) are two good examples of this; both were developed with little apparent regard to their long-term environmental effects – not even the well-known cautionary tale from an earlier decade, of the adverse effects of the indiscriminate use of DDT, appears to have made a lasting impression. It took two persistent scientists (Molina and Rowland 1974) to discover the ugly side of CFC gases, which was the fact that they had the propensity to destroy ozone, a fact now accepted as the prime cause for the depletion of the ozone layer in the outer atmosphere; and they have other adverse effects, yet commercially they promised so much good.

Air pollution results from the overburdening of the atmosphere with gases and chemicals. Like the oceans of the world, whose waters are able to assimilate and 'digest' quantities of pollutants in moderation, the air around us is able to absorb and dilute reasonable amounts of extra substances although a finite measurement cannot be placed on these quantities. Common sense must suggest that adverse effects result from the constant addition to the air of exhaust fumes from the smokestack of even one coal-burning power station.

The prime source of pollutant gases is from incineration. This may be natural as with the ejections from volcanoes or the fires of forest, bush and grassland; there is little we can do to prevent this. Within our control is the burning of fuel, mostly fossil fuel, for warmth and power. Either we have personal responsibility as with our open fires and the internal combustion engine in our car or motor cycle, or we may be indirectly responsible as with the centralized burning of coal, oil or gas necessary to produce the electricity available to us at the flick of a switch. We can recognize, and control, our personal contribution to pollution as we see smoke and fumes rise from our chimney, central heating flue or car exhaust pipe. What is not so obvious is that part of the smoke rising from the power station smokestack is ours as well.

The car manufacturer accurately named the three main culprit pollutants from car exhausts, namely nitrogen oxides (NO_x), hydrocarbons and carbon monoxide (CO). Sulphur dioxide (SO_2), the other main pollutant gas, is produced primarily when industry burns coal or other fossil fuels. Hydrogen, oxygen and carbon dioxide (CO_2) are non-toxic products of combustion, the two former usually in the form of water (H_2O).

Much has been made of lead as a pollutant, with the ensuing campaign for the use of unleaded petrol in order to clean our air. Lead is a health hazard, but since it is added to petrol in the first place, in order to ensure smooth-running engines, it is not a by-product of the combustion process itself. Lead-free air is essential, but to replace the fuel we use with unleaded petrol does not in any way resolve the problem of pollutant gases – in fact it tends to lull the general public into a false sense of security in relation to the poisons of vehicle exhaust. What it does prove is that, with the proper incentives and explanation, people are prepared to alter their habits.

Other pollutants, such as the CFCs and the PCBs, are now a part of our manufacturing processes and commercial or domestic life. Few people in the USA, UK or other western-style countries can be ignorant of the fact that aerosol containers, refrigerators and foam packaging are associated with gases harmful to the air. The 'ozone friendly' labelling on innumerable products reminds them of it daily. Again, perhaps, such an approach by industry and retailers brings about a feeling of complacency with regard to atmospheric pollution. The awful conditions, in some American cities, of the long, hot, dry summer of 1988 stimulated the *New York Times* (14 August 1988) to recognize that 'the suns and seas and sins of man have combined to transfer New York life into a seemingly endless slog through simmering broth'. This simmering broth was the pollution-laden air, thick with vehicle fumes and industrial haze.

Agriculture is not free of blame. The addition of artificial fertilizers to the land contributes mainly to an excess of nitrates and phosphates in the water runoff which enters the water cycle. But a considerable contribution is also made to the atmosphere, not least of all by the enormous increase in cattle rearing (over 1,000 million worldwide – McKibben 1990) and, to a lesser extent, other livestock. The digestive system which turns green fodder into food inside every beast also produces methane gas; 73 million tonnes (McKibben 1990) of this flatulence is expelled into the air and, globally, is a most significant contribution to the increase of methane in the atmosphere, up by 435 per cent over the last 100 years. Methane also derives from the process of decay. Rotting conditions as are found naturally in swamps are ideal for the generation of methane, which is often known as marsh or swamp gas. Paddy fields for growing rice imitate such conditions and provide a major source of methane. Well over 100 million cubic tonnes of the gas rise from the paddies annually. With the increase in world population, particularly in China, there needs to be an annual increase in rice acreage. Far from a decrease, we can only anticipate more methane from this source.

Termites can digest wood with bacteria in their intestinal tracts similar to those in cattle. They, too, excrete methane – every termite mound exudes 5 litres of the gas every minute. With an estimated half-tonne of termites for every person on Earth, some research scientists claim they make a considerable contribution to the release of this greenhouse gas (McKibben 1990). The dumping and rotting of domestic and commercial waste in pits leads on to a build-up of methane, dangerous directly because of its explosive nature and indirectly as an air pollutant. Methane is also a by-product of gas and petrol manufacture; collected from any source it is a valuable creator of power, yet only a few sporadic attempts are made to utilize it in the UK – for example, as at Packington Hall, Coventry, where waste disposal company PEEL exploits methane for electricity generation.

The interplay of natural forces provides various ways of counteracting pollution, that of air included. The process of photosynthesis is one of the most important whereby vegetation replaces carbon dioxide with oxygen. This vital link in the carbon cycle is able to provide a counterforce to the excessive production of CO_2, yet other activities of people, apart from that of direct pollution of the atmosphere, hinder this remedial action.

Timber extraction and other destruction of forested lands are a prime example, whilst the overloading of the seas with toxic materal is another. This is because the two major sources of plant material to perform the photosynthesis process grow in the tropical rain forests and on the oceans. In the former, masses of trees and other vegetation cover vast areas of land; with the latter, algal blooms extend across large expanses of the water surfaces. These are discussed in Chapter 6. It has been estimated that the CO_2 emitted from a 1,000 megawatt power station operating on 38 per cent thermal efficiency can be neutralized by the vegetation of 700 square miles of forest (McKibben 1990) (something like five days of rain forest destruction at the present rate). Environmental matters are full of contradictions – it is an ironic 'twist in the tail' of air pollution that some of the species of algae (*Phaeocystis pouchetti*, for example) are also a major source of SO_2, so that their CO_2-friendly activities are offset by their hostile contribution to acid rain.

So far my attempt has been to recognize the main pollutants and their sources, as well as to emphasize the holistic nature of the problems of air, and other, pollution. Mindful of the interrelation of these issues, let us look at the major problems individually.

The main problems associated with air pollution are:

- the contamination of the air we breathe;
- the creation of acid rain;
- global warming (the greenhouse effect);
- the depletion of the ozone layer.

CONTAMINATION OF THE AIR WE BREATHE

Clean air to breathe is an inalienable right of every person. It follows that it is equated with an individual responsibility not to foul the air. Yet the common concept of an ever improved level of living standards includes as major items:

- superior living accommodation;
- increased opportunity for travel (mainly through car ownership);
- the acquisition of goods and chattels;
- living and working in premises heated, cooled and managed by convenient services at the flick of a switch, the turn of a tap, the twist of a handle or the push of a button;
- better health facilities;
- increased opportunities to participate in leisure pursuits.

Most of these give rise, directly or indirectly, to the production of pollutant gases from the exhaust pipes of planes, trains, boats and road vehicles, the flues of factories and electricity power stations and the chimneys or extractor vents of homes, offices, shops, hospitals, leisure centres and public buildings.

The Worldwatch Institute (1990b) claims that a fifth of the world's population (about 1,000 million) is breathing air contaminated above international safety limits and concludes that tough international agreements are the only way to combat the spread of polluted air. There is some justification for the claim that

> air pollution, more than the existence of the Iron Curtain, brought about the revolution in Czechoslovakia. . . . The constant sore throats and headaches, the high incidence of lung disease and cancer, the lack of red blood corpuscles in the children, the low life expectancy were a nagging irritant – a daily, personal complaint against the old regime.
>
> (*Observer* 1990)

A crisis response to a recognized danger to pure air is not good enough. Long-term plans are required to eradicate the cause as opposed to short-term measures which alleviate the symptoms. This will be emphasized later, but a simple example will highlight the matter. The smog mask, commonly worn in Tokyo, Hamburg and Los Angeles, is finding favour with cyclists in Central London as individuals attempt to combat the inhalation of noxious fumes and particles of carcinogenous matter. These masks may help the wearer to avoid the worst of the evils in the air and they may help to advertise the existence of a problem, yet, at the same time, they may 'mask' the true problem as well, the problem of pollution caused by the inefficiency of the internal combustion engine and the incomplete incineration of a fossil fuel.

Encouragingly the full recognition of an air pollution problem backed by popular pressure to alleviate the cause can be effective. Smoky fog, which became known as 'smog', is one such example. It has been eradicated in Britain by the suppression of sooty particles, which poured from the chimneys of coal fires in susceptible areas. On days of temperature inversion when a layer of warm air sited itself above cooler air at ground level, the smog was trapped, and in London the 'pea-souper', as yellow as the lentils of the real soup, was created. A disastrous London smog episode in November 1952 caused the deaths of thousands of bronchitis sufferers and the complete disruption of the capital city for many days. The public uproar which followed led eventually to the Clean Air Act 1956 which set up legislation to impose smoke control zones in the UK and thus remove the smoke particles from the air in urban areas. Smokeless fuels, and a change to gas and electrical heating systems, have removed the dirt from the fumes, but not the gases. Now another 'smog' has appeared in some townscapes where nitrogen oxides from vehicle exhausts react with the sunlight to form ozone (O_3) and peroxy acetyl nitrate (PAN) in a photochemical process. Where a basin-shaped land area is host to thousands of petrol and diesel powered vehicles, the invisible spread of the gases gives rise to the photochemical smogs such as give notoriety to Los Angeles and Athens. By another ironic twist of nature, ozone, the very gas whose absence we deplore in the upper atmosphere, brings acute physical discomfort, and even more serious symptoms, to those who inhale it at ground level. PAN, the other main contaminant, apart from its minor effects, is thought to cause cancers; it also damages vegetation.

Vehicle emission reduction

Unlike the railway system, where the gross pollution of steam engines has given way to diesel oil powered locomotives and increasingly to electric trains, road vehicles remain firmly based on the petrol or oil fuelled internal combustion engine. Until a technological breakthrough occurs, such as the design of a ceramic engine to give a possible 200 miles per gallon fuel consumption, or the use of hydrogen as vehicle fuel (*Christian Science Monitor* 1990), there cannot be a fundamental improvement.

At present there are only two adaptations to the design of vehicles which can reduce, but not eliminate, pollutant emissions. The first has been mentioned already – the catalytic converter, already in general use in the USA and Japan on petrol driven vehicles. The three-way catalytic converter reduces the emission of CO, NO_x and hydrocarbons. It consists of an addition to the exhaust pipe similar in outward appearance to the silencer. Inside it contains a honeycomb structure coated with expensive metals such as platinum, palladium and rhodium. Reductions in pollutant gases of 90 per cent plus can be obtained with a conversion to nitrogen, water and CO_2. The life of a converter is limited to about 80,000 km (50,000 miles), which means it needs to be changed every three or more years depending on the use made of the vehicle. As catalysts the costly metals are not affected and can be recycled. It is essential to maintain the efficiency of the converter, which necessitates a national system of emission control, similar to the MOT testing at present. Lead damages the converter, so unleaded petrol must be readily available. It is also necessary to make certain that the engine is functioning well; this requires efficient combustion control. Even if the first owner of a car will accept the necessary restrictions it seems unlikely that subsequent owners, as the depreciation of the vehicle increases, will be so keen to pay out the necessary cost for replacement and maintenance. The other system is to use 'lean-burn' engines. As the petrol/air mixture is weakened the production of CO and hydrocarbons remains low while the NO_x falls away rapidly. Thus a lean-burn engine will not eliminate pollutants but it will reduce emissions for the whole life of the car. If a catalytic converter is used in conjunction with a lean-burn engine an even better result can be obtained. The question might be asked: if a silencer on cars is mandatory for UK vehicles, why not a pollution control converter?

There is something all drivers can do to lessen exhaust pollutants without design modifications. Slower speeds, especially if a constant level is maintained, reduce the creation of gases, and if this is coupled with the use of a fifth gear, which utilizes a slower engine revolution for the same road speed than a lower gear, then a minimum level of pollution is caused. Better still is to use a vehicle less for individual travel by making more use of public transport; this begs the question of an improved public system – available, reliable and regular.

The chemical industry

Direct pollution of the air from the chemical industry is well known through scenarios such as that of the leak of 40 tonnes of methyl isocyanate from the Union Carbide factory at Bhopal. Yet there is considerable evidence to show that the regular emission of toxic substances such as formaldehyde, hydrogen sulphide, silicon tetrafluoride and benzine has serious effects on the local population. At Abercwmboi, near Aberdare in south Wales, a smokeless fuel plant pours out those gases which would, without the prohibitions of smoke control zones, have poured from domestic chimneys elsewhere. A mere 20 miles way a Pontypool factory deals with toxic waste and denies the claim that it releases PCBs into the air. It is difficult to avoid the conclusion that the cost of remedial treatment to prevent the escape, or the complete destruction of pollutants, is only measured in financial terms and that the indirect costs to the welfare of people, animals and plants is not a consideration. Pearce, Markandya and Barbier in their *Sustainable Development, Resource Accounting and Project Appraisal: State of the Art Review* (1989), which was prepared as a report for the Department of the Environment, explore this economic aspect of environmental concern.

THE CREATION OF ACID RAIN

The term 'acid rain' has become the popular expression for any of the weather phenomena associated with excessive acidity. One of these matters is in fact the extra acidity associated with rainfall over and above that of the usual acidity of rain. Similarly, snow, hail and fog can all be over-acidic so that a better overall term is 'acid precipitation'. But acid rain does not preclude the inclusion of dry deposition of acidic particles, which in contact with dampness take on an acidic composition. Thus the preferred term for all of these acidic variations is 'acid deposition'.

Acidity is measured on the pH scale reflecting the concentration of hydrogen ions in the liquid. A pH of 7 is neutral, between 0 and 7 indicates that the substance is acidic, whilst a reading from 7 to 14 indicates an alkaline, or basic, state. The pH scale is logarithmic, so that between each point on the scale there is a tenfold difference. For rainfall which may measure 5.5 pH in normal circumstances to change to 3.5 pH may not sound alarming, but when this is interpreted to mean a 100-fold increase in acidity the magnitude of the variation becomes clear. Rainfall in the UK is normally acidic, between 5 and 6, because of the natural presence of carbon dioxide, and oxides of nitrogen and sulphur in the air. A rainfall reading of 2.4 (about the acidity of a lemon) was recorded at Pitlochry in Scotland on 20 April 1974 (Environment Canada 1982), and in late 1978 a pH of just under 2.0 was measured at Wheeling in West Virginia, USA (Eckholm 1982). Acid rain will not burn holes in cotton fabric – but even in its weak state it can destroy sensitive materials over a period of time.

The three 'common' acids of the atmosphere have been mentioned. Carbonic acid, nitric acid and sulphuric acid result from the chemical reaction of carbon gas, nitrogen gas and sulphur gas with atmospheric water. Of these the mixing of sulphur dioxide and rain leads to the most trouble unless the area affected happens to be a location where vehicular traffic is very heavy, when nitric acid, resulting from the reaction of nitrogen oxides with rain, will predominate. About 50 per cent of pollutant gases come from natural sources, but the other 50 per cent created by human activities are concentrated in fairly small areas relative to the size of the planet as a whole. This means that the major effects of acid rain are to be found in industrial concentrations such as the north-east USA, western Europe and the Far East.

Most of the pollutant gases come from the burning of fossil fuel as a power source. Natural gas, oil and coal (anthracite, steam coal, lignite or brown coal and peat) are the fossil fuels used to energize our industrial world. Their direct use is decreasing, for most are used to create electrical power. Coal in particular is sulphur rich, especially much of the coal mined in the UK. The release of this sulphur into the atmosphere as sulphur dioxide is the fundamental problem. The highest-level emitters of SO$_2$ in Britain are power stations – especially coal-burning ones – and refineries; next, as medium-level emitters, come other industries; low-level emitters are other non-domestic sectors (e.g. motor transport) and the domestic sector. To appreciate the immensity of the problem, observe the smoke rising from any of the thirty-eight coal-fired stations in the UK such as Ironbridge, Didcot, Fiddlers Ferry or Drax, and realize that this goes on every minute of the day and night, week after week, year after year.

The *emission* from a chimney is warmer than the air and thus rises into the atmosphere. Here the winds *transport* it away from the source, possibly across national borders, for it to react with water vapour and undergo *transformation* into an acidic mixture. This will fall as a *deposition* of acid rain on the land below, probably part of the territory of an 'innocent' nation as far as the origin of the pollutant is concerned. The speed at which this process occurs will vary according to the atmospheric and other conditions pertaining at the time.

The three main effects of acid deposition are on rivers and lakes, on trees and on buildings and other stone constructions.

Rivers and lakes

High acidity in a river or lake will lead to the death of most living things. Lakes, where the water is fairly static, are the most severely affected. The lake bed becomes covered with mosses and a few other growths which manage to exist. No algae are present to 'discolour' the water; no floating plants obscure the surface. The water is crystal clear, attractively enhanced by the reflection of sky, clouds and trees. It all looks most appealing, but the lake is dead – no fish, no invertebrates, few plants. The acidic rain leaches metals from the soil into the lake. Aluminium clogs the gills of fish, where such still exist, whilst the calcium in their skeletal structure is attacked and bones become brittle and unable to withstand the pull of muscle, resulting in deformity; eggs are infertile and what fry are produced are contorted in their shape. As the lake increases in acidity, specific creatures reach the ultimate level of their tolerance and die out. One indication, therefore, of the danger level reached is the 'suite' of plants and animals still in residence.

The time of greatest danger from acidity is in early spring when the snow is melting in the catchment areas which supply the water for the lake. Over the winter months snow will have fallen, bringing with it the acidity gained from pollutants in the upper air. As the ground cover of snow increases, the acid content will build up. Added to this there will be an evaporation of the snow surface as winter sunshine warms the top layer. Only the water will evaporate; the acid-causing chemicals will remain to increase the acidic ratio of chemical to snow. With the spring melt the water and the acidity will be released in a surge – an acid surge.

Sudden falls in pH values occur over very short periods of time; a 100-times increase in acidity over a twenty-day stretch is common (Figure 9.1). The effect on water life is devastating.

Trees

The leaching of chemicals from the soil has its deleterious effect on trees (Figure 9.2). Mineral deficiency is indicated by the premature discoloration and drop of deciduous leaves or coniferous needles. The direct attack of acid rain and acid mist adds to the harassment so that crown die-back is common and a thinning of the leaf canopy is apparent. Trees in a weakened state are more susceptible to permanent damage by attack from fungal disease and the ravages of ips, thrips and other insects. In particular trees growing on acidic soils are most sensitive to acid attack.

In high forest areas which are in close proximity to urban sprawls with dense vehicular traffic, the formation of photochemical ozone from the reaction of sunlight with nitrogenous exhausts will damage the trees. Such an area is the Black Forest of Germany and the hillsides, scarred with the defoliated trunks of dead trees, are testimony to the devastating effect of the pollutant gas.

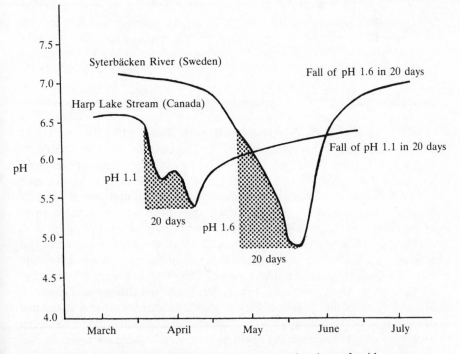

Figure 9.1 The increase in acidity in two rivers at the time of acid surge
Sources: Environment Canada; Stop Acid Rain Campaign, Sweden.

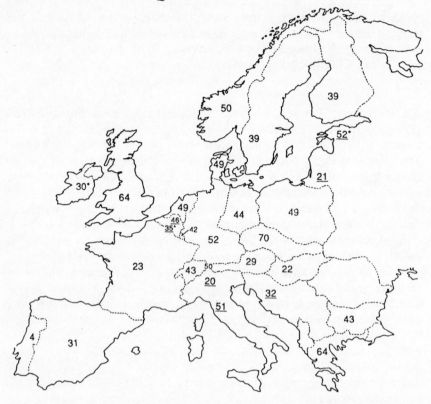

Figure 9.2 Intensity of defoliation in European countries, 1988: percentage of trees with more than 10 per cent defoliation
Sources: UN Economic Commission for Europe; International Co-operative Programme on Assessment and Monitoring of Air Pollution Effects on Forests.
Except when underlined (regional) the figures represent nationwide surveys
* = conifers only

Buildings

Dilute acids react with the lime in natural stone and concrete structures. Flakes of materials are lifted from the surface and the weakened features of decorative statuary crumble away. The ground below is littered with particles of eroded material. Measurements of the external structure of St Paul's Cathedral indicate that an average 30 mm of stone has disappeared from the outside walls in the last 250 years, whilst in some places 22 mm of it has crumbled away in the last 50 (Dudley *et al.* 1985). Stained glass is badly affected by acidity and many of the outstanding examples of ecclesiastical glass are fading rapidly; with some it has been necessary to sandwich the original between protective glass layers to halt further acid attack.

There is only one cure for acid rain and that is the elimination of pollutant gases at their source. It is almost certain that such is not possible in the foreseeable future as it will require a non-fossil-fuel technology as the basis for manufacturing industry and electrical power production. Even so it is possible to reduce the emissions of SO_2 from chimneys and the exhaust gases from internal combustion engines.

SO_2 removal

The sulphur is in the coal and oil before incineration. In the case of coal the sulphur content is between 0.5 and 5 per cent of its bulk, and with oil between 0.1 and 3 per cent (Parker and Trumbule 1987). Generally the soft coals in the UK have about 3 per cent sulphur content. The sulphur is contained in coal either as mineral grains of pyrites or as an integral chemical part of the fuel. It follows from this that it is relatively simple to remove the pyrites by washing the coal, allowing the heavier particles containing the sulphur to sink to the bottom. About a third of the sulphur could be removed in this way (Regens and Rycroft 1988). The removal of the chemical component is more difficult and therefore more expensive. It is hoped that commercially viable processes will come into operation in the early 1990s. One difficulty with such processes is that they create a large amount of solid waste. Compared with other forms of control the removal of sulphur before incineration is reliable and relatively inexpensive; it could be the best solution to the problem for developing countries.

The second method is to remove the gases at the burning stage. Most of the systems reduce both SO_2 and NO_x emissions, but they do require incorporation in the furnaces at the stage of factory construction, otherwise the complete change-over from one system to another is ultra-expensive. The two main methods are LIMB (lime injection in multistage burners) and FBC (fluidized bed combustion). Other systems are being developed, notably the IGCC (integrated coal gasification combined cycle). This method is used to create electric power with gas and steam turbines. Up to 99 per cent of the sulphur is removed and a great reduction in the amount of NO_x produced is made (Anderson 1987).

If fuel is used without remedial treatment and no combustion controls are made, the waste gases will be emitted from the smokestacks. Thus the only other way to deal with the pollutants is within the chimney system itself. FGD (flue gas desulphurization) employs methods which 'scrub' the exhaust gases. Again the commonest agent used is limestone in a wet or dry process. Ironically another method under development to treat the flue gases is an imitation of the acid rain process. In this E-beam (electron-beam) system both SO_2 and NO_x are removed by spraying the exhausts with water and ammonia. The resulting mixture passes through an electron-beam reactor which creates acids which in turn react with the ammonia and water to form compounds which can be used as fertilizers. Other E-beam systems are under consideration.

NO_x removal

NO_x produced in industrial technology has been considered above, for its prohibition by low-temperature incineration can be a part of the SO_2 removal system. Independent low-temperature furnaces can be installed to eradicate many of the NO_x problems from industry. But as far as acid rain is concerned the NO_x removal from vehicle exhaust is the prime consideration in the reduction of acids and ozone. This has already been described.

Liming

The cure for acid deposition has to be at source although something can be done to alleviate its effects where lakes, rivers and trees are concerned. There appears to be no economically viable technology which will prevent the reaction of acid rain with stonework, although where the object under attack is sufficiently valuable certain treatments can be employed. The introduction of lime into water can offset the effects of acidity. Putting lime on the surrounding land has the advantage of avoiding an alkaline surge – it is a more benevolent method which at the same time prevents the leaching of metals and other toxins from the soil. Liming is only palliative and not a cure; the cure has to be at source. An acid lake does not become a non-acidic lake – it becomes a lime-treated acidic lake and will revert without treatment.

As far as trees are concerned the liming process has to be evaluated over a long period of time. Unfortunately it may well be that the lime, by damaging soil organisms and leaching nitrates, actually harms tree growth. Liming the soil can at best maintain the adequate conditions until such time as acidification is reduced.

GLOBAL WARMING (THE GREENHOUSE EFFECT)[1]

The atmosphere becomes warmer as various gases absorb the long-wave radiation of heat from the Earth, in a way similar to the heating of the air in a greenhouse. The 'greenhouse effect' prevents warmth being dissipated into space and keeps the overall temperature of the Earth higher than it would be without these gases. Industrial activity over the past 150 years has increased the amount of CO_2 in the air by about a quarter and since this particular gas is the main 'greenhouse gas' it has resulted in a rise in temperature of 0.5°C. Over the next fifty years it is estimated that levels of CO_2 will rise by a further 30 per cent (Figure 9.3). There are many other greenhouse gases, notably methane, ozone,

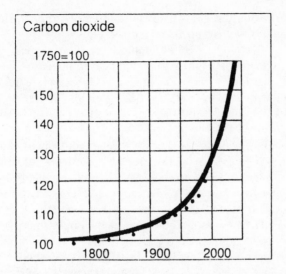

Figure 9.3 The rise in the level of CO_2 in the atmosphere over the last 250 years, taking 1750 to have a base level of 100
Source: UN Environment Programme.

nitrous oxide (laughing gas) and CFCs. Although the concentrations of these gases are much lower than that of CO_2, some of them have a greater greenhouse effect. Again over the next fifty years it is anticipated that these other gases will double the effect of CO_2. By making assumptions of the amount of each gas likely to be released into the air, computer models have predicted a potential rise in temperature of between 1.5 and 4.5°C by 2050. The oceans of the world will take longer to warm, so that the increase will be spread over a relatively long period.

These temperature rises may seem to be small and of little real significance. However, the warming will not be spread evenly, and local variations may see increases of more than 10°C at the polar regions.

What then will be the effect of this global warming? When heated, water expands and occupies more space. This will be the prime cause of higher sea levels, although increased melting of the south polar ice cap will add to this. Since many of the major towns of the world are located by the sea, the effect may be catastrophic. Certainly it is doubtful if the recently completed Thames Barrage will be high enough to cope with the spring tides rising above a higher initial base level. It is generally agreed that larger temperature changes occurring in high latitudes will affect the grain-growing areas particularly. Marginal lands will be at risk and the creation of another Dust Bowl in the USA is more than an unpleasant thought. Even if warmer temperatures speed crop growth and increase yields (extra CO_2 has a similar effect) they will also benefit pests, weeds and diseases. Increased rainfall will lower wheat production in the most northerly regions as, for example, on the Canadian prairies – some estimates say by up to a quarter.

It is likely that a change of climate in already marginal areas of farming, such as the Sahel in Africa, will lead to increased desertification, drought and erosion of the soil. In our own temperate region, winters may become warmer and wetter, with drier and hotter summers. This assessment is complicated by the effect of cloud cover, which is almost impossible to predict.

If the greenhouse effect warms the oceans and thaws the permafrost of the tundra regions it will aid the release of the 10 million million tonnes of methane estimated to be 'locked' in the muds of the continental shelves and the frozen soils. A release of a potential half-million tonnes annually would probably double the present concentration of methane in the atmosphere. A dangerous 'feedback loop' would be created: a warmer atmosphere releases more methane – more methane further warms the atmosphere (McKibben 1990).

It is perhaps too late to prevent global warming.

A child born now will never know a natural summer, a natural autumn, winter or spring. Summer is becoming extinct, replaced by something else which will be called 'summer'. This new summer will retain some of its relative characteristics – it will be hotter than the rest of the year for instance, and will be the time of the year when crops grow but it will not be summer, just as the best prosthesis is not a leg.

(McKibben 1990: 55)

Yet it is not too late to do all that can be done to minimize the greenhouse effect and to reduce its causes. The controls needed to reduce CO_2 and the other gases have been discussed already, although the present trend to improve the quality of vehicle exhaust through the use of catalytic converters does little to prevent CO_2 emissions. An overall policy which includes reduction in the incineration of fossil fuels by domestic and industrial activities, more efficient and economical use of internal combustion engines, control on the production and use of CFCs and a reduction in deforestation must be matched by a world policy of energy conservation. As the developing nations of the world anticipate a rise in living standards, it is essential that they are enabled to make technological advances which avoid the need to increase the pollution of the atmosphere – this will take massive financial aid, hopefully available from defence budgets in a 'friendlier' world.

DEPLETION OF THE OZONE LAYER

Ozone, composed of three atoms of oxygen, is concentrated in a zone between 20 and 25 km above the Earth. It has been said to protect the planet and its people like a delicate veil protects the face beneath. Solar radiation contains ultraviolet rays and of these UV-B causes skin cancer, ageing and wrinkling, and eye malfunctions in humans, slows plant growth, destroys marine algae and fish larvae and breaks down the chemical structure of paints and plastics. At our present levels of UV-B white paints change to yellow, coloured pigments fade and fabrics rot under its influence. Any depletion in the amount of O_3 increases the UV radiation which reaches the Earth. Sun-bathing by those with light-coloured skins, already a potentially harmful activity, may become a risk to health to rank alongside smoking, drug and alcohol abuse. Estimates indicate that between 10 and 30 per cent of the sun's UV-B reaches the surface of the Earth at the present time. Were O_3 levels to fall by only 10 per cent the increase in UV-B would be 20 per cent. In the USA the National Academy of Sciences has estimated that every 3 per cent reduction in O_3 would result in 20,000 more skin cancer sufferers annually in that country alone (Watson *et al.* 1986).

The British Antarctic Survey has been at the forefront of ozone layer investigation. It was the scientists of this organization who first recognized the serious depletion of ozone in the polar region, which, during the Antarctic spring, resulted in a 'hole' in the layer. A similar depletion in the layer above the Arctic has also been recorded (Figure 9.4).

O_3 is an unstable form of oxygen, readily giving up the third atom, particularly to atoms of chlorine. CFC gases are not destroyed by the usual chemical reactions in the lower atmosphere. Instead they rise into the upper atmosphere where ultraviolet radiation causes free chlorine atoms to be released. These collect one of the oxygen atoms to form chlorine monoxide and oxygen. A further reaction releases the chlorine atom, thus again freeing it to destroy ozone in a 'chain' sequence (Figure 9.5). Nitrous oxide also plays its part in O_3 destruction and its release from high-flying aircraft is a contributory factor.

Fortunately the danger of the depletion of the ozone layer has now

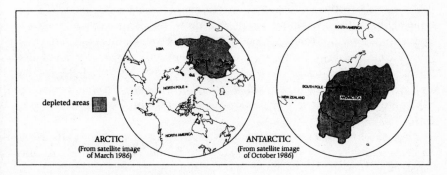

Figure 9.4 Thinning of the polar ozone layer
Source: Environment Canada.

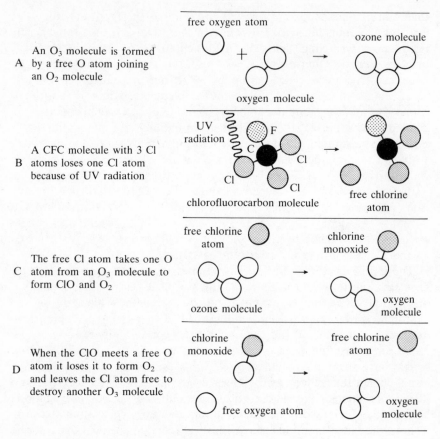

A   An O₃ molecule is formed by a free O atom joining an O₂ molecule

B   A CFC molecule with 3 Cl atoms loses one Cl atom because of UV radiation

C   The free Cl atom takes one O atom from an O₃ molecule to form ClO and O₂

D   When the ClO meets a free O atom it loses it to form O₂ and leaves the Cl atom free to destroy another O₃ molecule

Figure 9.5 How CFCs destroy ozone
Source: UN Environment Programme.

been recognized internationally and many conferences have been held on the subject. Arguably that held in Montreal, Canada, in September 1987 was the most significant as it stimulated the debate on CFCs. A change in attitude to 'green' matters caused the British government to convene an ozone layer conference (March 1989) which brought together countries from most of the world. The Montreal Protocol demanded a 50 per cent reduction in the use of CFCs by 1999 – in the light of new evidence this was accepted as inadequate in London; Britain, for example, pledged itself to bring down its CFC emission by 85 per cent by 1990.

The use of CFCs as aerosol propellants and in refrigeration units needs to cease completely and the gas already in redundant refrigeration units must be prevented from escape into the air. Alternative foaming agents must be used. Generally it seems that HFCs (hydrofluorocarbons) may be the answer in the short term, but in due time the use of HFAs (hydrofluoroalkanes) should be a better alternative (ICI 1990). As for the use of CFCs in the electrical industry, tighter production controls can lead to a full recycling of the gases to prevent their release into the air.

CONCLUSION

The movement of air, the wind, ensures that atmsopheric pollution cannot be considered in isolation.

> A few countries have so far accounted disproportionately for the rise in carbon dioxide but . . . the resulting climatic changes will not be allocated according to any earthly idea of justice. No nation acting alone can prevent an increase in atmospheric carbon dioxide.
>
> (Eckholm 1982)

If for CO_2 any of the pollutant gases is substituted the feelings expressed here are still valid. Air pollution may be seen to consist of individual problems, but they are interrelated, and preventive measures are a global problem.

Mostafa Tolba, the executive director of UNEP, speaking at the UN conference on global warming at Villach, Austria, in 1985 (World Meteorological Organization 1986), emphasized that 'In the 300 years or so that have encompassed the agricultural and industrial revolutions, man has begun to replace nature as the engine of climatic change.' It follows from this that some sort of control of this and other air pollution factors can be made if international will is positively inclined. The 'will' has to include the compliance of developed nations to underwrite the cost of the more expensive 'environmentally friendly' technology. This will enable the developing nations to exploit their natural reserves of fossil fuel or alternative sources of power production, in order to raise the living standards of their populations. The accumulation of toxic material in the air must be kept to a minimum and concentrations of solid, liquid and gaseous matter resulting from human activities must not be allowed to reach levels where they interfere with the natural cycles of a healthy atmosphere.

It is not only 'nations' which have a responsibility to change their habits and practices – it is a personal responsibility which centres primarily on energy conservation, less reliance on individual engine powered transport and a diminution of demand for manufactured goods. These need to be matched with an awareness that political pressure is necessary to bring about a true economic assessment of 'development', an assessment which places a value other than cultural on environmental loss and atmospheric pollution. Air pollution and other environmental problems require international financial intervention which will enable governments, especially of the Third World, to tackle pollution problems at source and to introduce non-polluting technology. After the Second World War the USA channelled more assistance per US citizen to the war-shattered economies of Europe than the entire world now spends on development assistance.

'I am advocating support for a meaningful international environmental fund. Not of millions, but of billions of dollars,' said Mostafa Tolba in a speech in Brussels on World Environment Day 1989 (UNEP 1989a). 'This is not a day-dream. It can be done.' To which I would add: It must be done.

NOTE

1 Statistics on the greenhouse effect show wide variations which reflect different climatic models. Those used here are provided by the United Nations Environment Programme (UNEP 1987).

10 Challenges in environmental ethics

Holmes Rolston III

Ethicists had settled on at least one conclusion as ethics became modern in Darwin's century: that the moral has nothing to do with the natural. Science describes natural history and natural law; ethics prescribes human conduct, moral law; and to confuse the two makes a category mistake, commits the naturalistic fallacy.[1] Nature simply *is*, without objective value; the preferences of human subjects establish value; and these human values, appropriately considered, generate what *ought* to be. Nature is amoral; only humans are ethical subjects and objects of duty.

Unsettled now, in the decade in which we prepare to enter the next millennium, ethicists face foreboding revolution. Only the human species contains moral agents, but perhaps conscience on Earth ought not be used to exempt every other form of life from consideration, with the resulting paradox that the sole moral species acts only in its collective self-interest towards all the rest. There is something overspecialized about an ethic, held by the dominant class of *Homo sapiens*, that regards the welfare of only one of several million species as an object and beneficiary of duty. Without denying the duties within culture, this biological world that *is* also *ought to be*; we must argue from the natural to the moral.

There is something Newtonian, not yet Einsteinian, besides something morally naive, about living in a reference frame where one species takes itself as absolute and values everything else relative to its utility. If true to their specific epithet, ought not *Homo sapiens* value this host of life as something with a claim to care in its own right? Man may be the only measurer of things, but is man the only measure of things? The challenge of environmental ethics is a principled attempt to redefine the boundaries of ethical obligation.

Faced with revolution, ethical conservatives may shrink back and refuse to think biologically, to naturalize ethics in the deep sense. They will fear that it is logically incoherent to suppose there is non-anthropogenic value, or that this is too metaphysically speculative ever to be operational. They will insist that it does not make any pragmatic difference anyway, claiming that an adequate environmental ethic can be anthropogenic, even anthropocentric.

When we face up to the crisis, however, we undergo a more direct moral encounter. Environmental ethics is not a muddle; it is an invitation to moral development. All ethics seeks an appropriate respect for life, but respect for human life is only a subset of respect for all life. What ethics is about, in the end, is seeing outside your own sector of self-interest, of class interest. A comprehensive ethic will find values in and duties to the natural world. An ecological conscience requires an unprecedented mix of science and conscience, of biology and ethics.

RESPECT FOR PLANTS AND ANIMALS

We have direct encounters with life that has eyes, at least where behind those eyes there seems to reside a subject of experience. So it is not difficult to begin to extend an ethic to mammals, to warm-blooded animals, even to all vertebrates. We sense that there is somebody there behind the fur and feathers, somebody there whose pains and pleasures count to them, and ought to us. We should treat animals, as we rather curiously say, 'humanely', meaning that so far as they, cousins to our human selves, can suffer, they are morally considerable. Just how much they should count remains unsettled (ought they to be eaten?) but to reach the conclusion that their 'welfare' or 'rights' should count at all begins to move us out of an exclusively anthropic into an environmental ethic. But we have only begun to mix biology and ethics, and the principal challenges lie ahead.

In college zoology I did an experiment on nutrition in rats, to see how they grew with and without vitamins. When the experiment was completed, I was told to take the rats out and drown them. I felt squeamish but did it. In college botany I did an experiment on seedlings to test how they grew with this or that fertilizer. The experiment over, I threw out the seedlings without a second thought. While there can be ethics about sentient animals, after that perhaps ethics is over. Respect for life ends somewhere in zoology; it is no part of botany. No consciousness, no conscience. Without sentience, ethics is nonsense.

Or do we want an ethic that is more objective about life? In Yosemite National Park for almost a century humans entertained themselves by driving through a tunnel cut in a giant sequoia. Two decades ago the Wawona tree, weakened by the cut, blew down in a storm. People said: Cut us another drive-through sequoia! The Yosemite environmental ethic, deepening over the years, said: No! You ought not to mutilate majestic sequoias for amusement. Indeed, some ethicists count the value of redwoods so highly that they will spike redwoods, lest they be cut. In the Rawah Wilderness in alpine Colorado, old signs read: 'Please leave the flowers for others to enjoy.' When they rotted out, the new signs urged a less humanist ethic: 'Let the flowers live!'

But trees and flowers cannot care, so why should we? We are not considering animals that are close kin, nor can they suffer or experience anything. There are no humane societies for plants. Plants are not valuers with preferences that can be satisfied or frustrated. It seems odd to claim that plants need our sympathy, odd to ask that we should consider their point of view. They have no subjective life, only objective life.

Fishermen in Atlantic coastal estuaries toss beer bottles overboard, a convenient way to dispose of trash. On the bottom, small crabs, attracted by the residual beer, make their way inside and become trapped, unable to get enough foothold on the slick glass neck to work their way out. They starve slowly. Then one dead crab becomes bait for the next victim, an indefinitely resetting trap. Are those bottle traps of ethical concern? Or is the whole thing out of sight, out of mind, with crabs too mindless to care about? Should sensitive fisherman pack their bottle trash back to shore – whether or not crabs have much, or any, felt experience?

Flowers and sequoias live; they ought to live. Crabs have value out of sight, out of mind. Afraid of the naturalistic fallacy, conservative ethicists will say that people should enjoy letting flowers live or that it is silly to cut drive-through sequoias, aesthetically more excellent for humans to appreciate both for what they are. The crabs are out of sight, but not really out of mind; humans value them at a distance. But these ethically conservative reasons really do not understand what biological conservation is in the deepest sense. Nothing matters to a tree, but much is *vital*.

An organism is a spontaneous, self-maintaining system, sustaining and reproducing itself, executing its programme, making a way through the world. It can reckon with vicissitudes, opportunities and adversities that the world presents. Something more than phyical causes, even when less than sentience, is operating within every organism. There is *information* superintending the causes; without it the organism would collapse into a sand heap. This information is a modern equivalent of what Aristotle called formal and final causes; it gives the organism a *telos*, 'end', a kind of (non-felt) goal.[2]

This information is coded in the DNA; the genome is a set of *conservation* molecules. Given a chance, these molecules seek organic self-expression. They thus proclaim a life way, and with this an organism, unlike an inert rock, claims the environment as source and sink, from which to abstract energy and materials and into which to excrete them. Life thus arises out of earthen sources (as do rocks), but life turns back on its sources to make resources out of them (unlike rocks). An acorn becomes an oak; the oak stands on its own.

So far we have only description. We begin to pass to value when we recognize that the genetic set is a *normative set*; it distinguishes between what *is* and what *ought to be*. This does not mean that the organism is a moral system, for there are no moral agents in nature; but the organism is an axiological, evaluative system. So the oak grows, reproduces, repairs its wounds and resists death. The physical state that the organism seeks, idealized in its programmatic form, is a valued state. *Value* is present in this achievement. *Vital* seems a better word for it than *biological*.

A life is defended for what it is in itself, without necessary further contributory reference, although, given the structure of all ecosystems, such lives necessarily do have further reference. The organism has something it is conserving, its life. Organisms have their own standards, fit into their niche though they must. They promote their own realization, at the same time that they track an environment. Every organism has a *good-of-its-kind*; it defends its own kind as a *good kind*. In that sense, as soon as one knows what a giant sequoia tree is, one knows the biological identity that is sought and conserved. Man is neither the measurer nor the measure of things; value is not anthropogenic, it is biogenic. A moral agent deciding his or her behaviour ought to take account of the consequences for other evaluative systems.

That will not be our only consideration, because organisms, each with a *good-of-its-kind*, may well not be *good-in-their-place*, may not be good in their ecosystems or good for satisfying human preferences. Organisms – we may expect from ecological theory – will for the most part be adapted fits in their ecosystems; their genetic programmes will have to mesh with their ecosystemic roles. So it is likely that in wild nature there will be some selection pressures for a situated environmental fitness. Not all such situations are good arrangements. The vicissitudes of historical evolution sometimes result in ecological webs that are sub-optimal solutions, within the biologically limited possibilities and powers of interacting organisms.

Still, such systems have been selected over millennia for functional stability; and at least the burden of proof is on a human evaluator to say that any natural kind is a bad kind and ought not to call forth admiring respect. These claims about good kinds do not say that things are perfect kinds, or that there can be no better ones, only that natural kinds are good kinds until proven otherwise.

What is almost invariably meant by a 'bad' kind is that an organism is instrumentally bad when judged from the viewpoint of human interests, of humane interests. 'Bad' so used is an anthropocentric word; there is nothing at all biological or ecological about it, and so it has no force evaluating objective nature, however much humanist force it may sometimes have.

A really *vital* ethic respects all life, not just animal pains and pleasures, much less just human preferences. In the Rawahs, the old signs, 'Leave the flowers for others to enjoy', were application signs using an old, ethically conservative, that is, *humanistic*, ethic. The new ones invite a change of reference frame – a wilder, more logical because more biological ethic, a radical ethic that goes down to the roots of life, that really is conservative because it understands *biological* conservation at depths.

What the injunction 'Let the flowers live!' means is: 'Daisies, marsh-marigolds, geraniums, larkspurs are evaluative systems that conserve goods of their kind and, in the absence of evidence to the contrary, are good kinds. Is there any reason why your human interests should not also conserve these good kinds?' A drive-through sequoia causes no suffering; it is not cruel. But it is callous and insensitive to the wonder of life. Some will complain that we have committed the naturalistic fallacy; rather, we invite a radical commitment to respect all life.

RESPECT FOR ENDANGERED SPECIES

Certain rare species of butterflies occur in hummocks (slightly elevated forested ground) on the African grasslands. It was formerly the practice of unscrupulous collectors to go in, collect a few hundred specimens and then burn out the hummock with the intention of destroying the species, thereby driving up the price of their collections. I find myself persuaded that they morally ought not do this. Nor will the reason resolve into the evil of greed, but it remains the needless destruction of a butterfly species.

This conviction remains even when the human goods are more worthy. The Bay checkerspot, *Euphydryas editha bayensis*, is endemic to the San Francisco Bay region. It is proposed to be listed as an endangered species and inhabits peripheral tracts of a large facility on which United Technologies Corporation, a missile contractor, builds and tests Minute-man and Tomahawk propulsion systems. The giant defence contractor has challenged the proposed listing, thinks it airy and frivolous that a butterfly should slow the delivery of warhead missile propulsion systems and so went ahead and dug a water pipeline through a butterfly patch. They operated out of the classical ethics that says that butterflies do not count but that the defence of humans does. But a more radical, environmental ethics demurs. The good of humans might override the good of butterfly species but the case must be argued.

A species exists; a species ought to exist. Environmental ethics must make both claims and move from biology to ethics with care. Do species exist? Species exist only instantiated in individuals, yet species are as real as individual plants or animals. The claim that there are specific forms of life historically maintained in their environments over time seems as certain as anything else we believe about the empirical world. At times biologists revise the theories and taxa with which they map these forms, but species are not so much like lines of latitude and longitude as like mountains and rivers, phenomena objectively there to be mapped. One species will slide into another over evolutionary time. But it does not follow from the fact that speciation is sometimes in progress that species are merely made up, not found as evolutionary lines with identity in time as well as space.

Ought species to exist? A consideration of species is revealing and challenging because it offers a biologically based counter-example to the focus on individuals – typically sentient and usually persons – so characteristic in classical ethics. In an evolutionary ecosystem, it is not mere individuality that counts, but the species is also significant because it is a dynamic life form maintained over time. The individual is a token of a type, and the type is more important than the token.

A species lacks moral agency, reflective self-awareness, sentience or organic individuality. Some will be tempted to say that specific-level processes cannot count morally. Duties must attach to singular lives, most evidently those with a psychological self, or some analogue to this. In an individual organism, the organs report to a centre; the good of a whole is defended. The members of a species report to no centre. A species has no self. There is no analogue to the nervous hook-ups or circulatory flows that characterize the organism.

But singularity, centredness, selfhood, individuality, are not the only processes to which duty attaches. A more radical ethic knows that having a biological identity reasserted genetically over time is as true of the species as of the individual. Identity need not attach solely to the centred organism; it can persist as a discrete pattern over time. Thinking this way, the life that the individual has is something passing through the individual as much as something it intrinsically possesses. The individual is subordinate to the species, not the other way around. The genetic set, in which is coded the *telos*, is as evidently the property of the species as of the individual through which it passes. A consideration of species strains any ethic fixed on individual organisms, much less on sentience or persons. But the result can be biologically sounder, though it revises what was formerly thought logically permissible or ethically binding.

The species line is the *vital* living system, the whole, of which individual organisms are the essential parts. The species too has its integrity, its individuality, its 'right to life' (if we must use the rhetoric of rights); and it is more important to protect this vitality than to protect individual integrity. The right to life, biologically speaking, is an adaptive fit that is right for life, that survives over millennia, and this generates at least a presumption that species in niche are good right where they are, and therefore that it is right for humans to let them be, to let them evolve. The integrity resides in the dynamic form; the individual inherits this, exemplifies it and passes it on. The appropriate survival unit is the appropriate level of moral concern.

Sensitivity to this level, however, can sometimes make an environmental ethicist seem callous. On San Clemente Island, the US Fish and Wildlife Service and the California Department of Fish and Game planned to shoot two thousand feral goats to save three endangered plant species, *Malacothamnus clementinus*, *Castilleja grisea*, *Delphinium kinkiense*, of which the surviving individuals numbered only a few dozens.

After a protest, some goats were trapped and relocated. But trapping all was impossible and many hundreds were killed. Is it inhumane to count plant species more than mammal lives, a few plants more than a thousand goats?

Those who wish to return rare species of big cats to the wild have asked about killing genetically inbred, inferior cats, presently held in zoos, in order to make space available for the cats needed to reconstruct a population genetically more likely to survive upon release. All the Siberian tigers in zoos in North America are descendants of seven animals; if these were replaced by others nearer to the wild type with more genetic variability, the species could be saved in the wild. When we move to the level of species, we may kill individuals for the good of their kind.

Humans have more understanding than ever of the natural world, of the speciating processes, more predictive power to foresee the intended and unintended results of their actions and more power to reverse the undesirable consequences. The duties that such power and vision generate no longer attach simply to individuals or persons but are emerging duties to specific forms of life. The wrong that humans are doing, or allowing to happen through carelessness, is stopping the historical vitality of life, the flow of natural kinds.

Every extinction is an incremental decay in this stopping life, no small thing. Every extinction is a kind of super-killing. It kills forms (*species*), beyond individuals. It kills 'essences' beyond 'existences', the 'soul' as well as the 'body'. It kills birth as well as death. Afterwards nothing of that kind either lives or dies. A shut-down of the life stream is the most destructive event possible. Never before has this level of question – super-killing by a super-killer – been deliberately faced. What is ethically callous is the maelstrom of killing and insensitivity to forms of life and the sources producing them. What is required is principled responsibility to the biospheric Earth.

Several billion years' worth of creative toil, several million species of teeming life, have been handed over to the care of this late-coming species in which mind has flowered and morals have emerged. Life on Earth is a many splendoured thing; extinction dims its lustre. If, in this world of uncertain moral convictions, it makes any sense to claim that one ought not to kill individuals, without justification, it makes more sense to claim that one ought not to super-kill the species, without super-justification. That moves from what *is* to what *ought to be*. But this is no fallacy being committed by ethical naturalists, rather it is the humanists who cannot draw these ethical conclusions who are mistaken in their logic.

RESPECT FOR ECOSYSTEMS

'A thing is right,' urged Aldo Leopold, concluding his land ethic, 'when it tends to preserve the stability and integrity of the biotic community; it is wrong when it tends otherwise' (Leopold 1949: 224–5). Again, we have two parts to the ethic: first, that ecosystems exist, both in the wild and in support of culture; second, that ecosystems ought to exist, both for what they are in themselves and as modified by culture. Again, we must move with care from the biological to the ethical claims.

Classical humanistic ethics finds ecosystems to be unfamiliar territory. It is difficult to get the biology right and, superimposed on the biology, to get the ethics right. Fortunately, it is often evident that human welfare depends on ecosystemic support, and in this sense all our legislation about clean air, clean water, soil conservation, forest policy, pollution controls, oil spills, renewable resources and so forth is concerned about ecosystem-level processes. Pragmatically, we can often decide what to do while the full set of our reasons is still being discussed.

Still, a comprehensive environmental ethics needs the best, naturalistic reasons, as well as the good, humanistic ones, for respecting ecosystems. When we have these reasons, we may also think that we ought to restore wolves for the integrity of ecosystems, or to let natural fires burn, or to remove feral animals that are degrading the system, or to set aside wilderness areas untrammelled by humans, or to restore prairie ecosystems. We may refuse to stock game fish if these imperil riparian ecosystems, or prohibit an apartment complex in order to save a swamp, or refuse to drill for oil in a pristine ecosystem, or insist on double-hulled tankers lest a marine ecosystem be put at needless risk.

The ecosystem is the community of life; in it the fauna and flora, the species, have entwined destinies. Ecosystems generate and support life, keep selection pressures high, enrich situated fitness, evolve diverse and complex kinds. The ecologist finds that ecosystems are objectively satisfactory communities in the sense that organismic needs are sufficiently met for species long to survive, and the ethicist finds (in a subjective judgement matching the objective process) that such ecosystems are satisfactory communities to which to attach duty. Our concern must be for the fundamental unit of survival.

An ecosystem, a traditional ethicist may say, is too low a level of organization to be respected intrinsically. Ecosystems can seem little more than random, statistical processes. A forest can seem a loose collection of externally related parts, the collection of fauna and flora a jumble, hardly a community. The plants and animals within an ecosystem have needs, but their interplay can seem simply a matter of distribution and abundance, birth rates and death rates, population densities, parasitism and predation, dispersion, checks and balances, stochastic process. There is only catch-as-catch-can scrimmage for nutrients and energy, a game played with loaded dice, not really enough integrated process to call the whole a community.

Unlike higher animals, ecosystems have no experiences; they do not and cannot care. Unlike plants, an ecosystem has no organized centre, no genome. It does not defend itself against injury or death. Unlike a species, there is no ongoing *telos*, no biological identity reinstantiated over time. The organismic parts are more complex than the community whole. More troublesome still, an ecosystem can seem a jungle where the fittest survive, a place of contest and conflict, beside which the organism is a model of co-operation. It can seem that only organisms are 'real', actually existing as entities, whereas ecosystems are nominal – just interacting individuals. Oak trees are real, but forests are nothing but collections of trees.

But any level is real if it shapes behaviour on the level below it. Thus the cell is real because that pattern shapes the behaviour of amino acids; the organism because that pattern co-ordinates the behaviour of hearts and lungs. The biotic community is real because the niche shapes the morphology of the oak trees within it. Being real at the level of community only requires an organization that shapes the behaviour of its members.

The challenge is to find a clear model of community and to discover an ethics for it – better biology for better ethics. Even before the rise of ecology, biologists began to conclude that the combative survival of the fittest distorts the truth. The more perceptive model is coaction in adapted fit. Predator and prey, parasite and host, grazer and grazed are contending forces in dynamic process where the well-being of each is bound up with the other – co-ordinated as much as heart and liver are co-ordinated organically. The ecosystem supplies the co-ordinates through which each organism moves, outside which the species cannot really be located. A species is what it is where it is.

The community connections are looser than the organism's internal interconnections – but not less significant. Admiring organic unity in organisms and stumbling over environmental looseness is like valuing mountains and despising valleys. Internal complexity – heart, liver, muscles, brain – arises as a way of dealing with a complex, tricky, but challenging environment. The skin-out processes are not just the support, they are the subtle source of the skin-in processes. In the complete picture, the outside is as vital as the inside. Had there been no ecosystemic community, no organismic unity could have evolved. There would be less elegance in life.

An ecosystem is a productive, projective system. Organisms defend only their selves, with individuals defending their continuing survival and species increasing the numbers of kinds. But the evolutionary ecosystem spins a bigger story, limiting each kind, entwining it with the welfare of others, promoting new arrivals, bringing forth kinds and the integration of kinds. Species *increase their kind*; but ecosystems *increase kinds*, superposing the latter increase onto the former. *Ecosystems are selective systems, as surely as organisms are selective systems.* The individual is programmed to make more of its kind, but more is going on systemically than that; the system is making more kinds.

Hence the evolutionary toil, elaborating and diversifying the biota, that once began with no species and results today in 5 million species, increasing over time the quality of lives in the upper rungs of the tropic pyramids. One-celled organisms evolved into many-celled, highly integrated organisms. Photosynthesis evolved and came to support locomotion – swimming, walking, running, flight. Stimulus-response mechanisms became complex instinctive acts. Warm-blooded animals followed cold-blooded ones. Complex nervous systems, conditioned behaviour and learning emerged. Sentience appeared – sight, hearing, smell, tastes, pleasure, pain. Brains coupled with hands. Consciousness and self-consciousness arose. Culture was superposed on nature.

These developments do not take place in all ecosystems or at every level. Microbes, plants and lower animals remain, good of their kinds and serving continuing roles, good for other kinds. The understoreys remain occupied. As a result, the quantity of life and its diverse qualities continue – from protozoans to primates to people. The later we go in time the more accelerated are the forms at the top of the tropic pyramids, the more elaborated are the multiple tropic pyramids of Earth. There are upward arrows over evolutionary time. An ecosystem has no head, but it has a 'heading' for species diversification, support and richness. Though not a super-organism, it is a kind of vital field.

Instrumental value uses something as a means to an end; *intrinsic value* is worthwhile in itself. No warbler eats insects to become food for a falcon; the warbler defends its own life as an end in itself and makes more warblers as she can. A life is defended intrinsically, without further contributory reference. But neither of these traditional terms is satisfactory at the level of the ecosystem. Though it has value *in* itself, the system does not have any value *for* itself. Though a value producer, it is not a value owner. We are no longer confronting instrumental value, as though the system were of value instrumentally as a fountain of life. Nor is the question one of intrinsic value, as though the system defended some unified form of life for itself. We have reached something for which we need a third term: *systemic value*. Duties arise in encounter with the system that projects and protects these member components in biotic community.

Ethical humanists will say that ecosystems are of value only because they contribute to human experiences. But that mistakes the last chapter for the whole story, one fruit for the whole plant. Humans count enough to have the right to flourish on Earth, but not so much that they have the right to degrade or shut down ecosystems, not at least without a burden of proof that there is an overriding cultural gain. The Environmental ethics does say that ecosystems are of value because they contribute to human welfare, to animal experiences and to plant life, but it couples that with the more radical view that the stability, integrity and beauty of biotic communities are what are most fundamentally to be conserved.

In environmental ethics one's beliefs about nature, which are based upon but exceed science, have everything to do with beliefs about duty. The way the world *is* informs the way it *ought* to be. We always shape our values in significant measure in accord with our notion of the kind of universe that we live in, and this drives our sense of duty. Our model of reality implies a model of conduct. Perhaps we can leave open what metaphysics ultimately underlies our cosmos, but for an environmental ethics at least we will need an Earthbound metaphysics, a meta-ecology. Differing models sometimes imply similar conduct, but often they do not. A model in which nature has no value apart from human preferences will imply different conduct from one where nature projects fundamental values, some objective and others that further permit human subjective experiences superposed on objective nature.

This evaluation is not scientific description; hence not ecology *per se*, but we do move to meta-ecology. No amount of research can verify that, environmentally, the right is the optimum biotic community. Yet ecological description generates this valuing of nature, endorsing the systemic rightness. The transition from *is* to *good* and thence to *ought* occurs here; we leave science to enter the domain of evaluation, from which an ethic follows.

What is ethically puzzling and exciting is that an *ought* is not so much *derived* from an *is* as discovered simultaneously with it. As we progress from descriptions of fauna and flora, of cycles and pyramids, of autotrophs co-ordinated with heterotrophs, of stability and dynamism, on to intricacy, planetary opulence and interdependence, to unity and harmony with oppositions in counterpoint and synthesis, organisms evolved within and satisfactorily fitting their communities, arriving at length at beauty and goodness, it is difficult to say where the natural facts leave off and where the natural values appear. For some at least, the sharp *is/ought* dichotomy is gone; the values seem to be there as soon as the facts are fully in, and both alike properties of the system. This conviction, and the conscience that follows from it, can yield our best adaptive fit on Earth.

NOTES

1 The expression 'naturalistic fallacy' was coined by G. E. Moore (1903) for the attempt to define the concept of Good – which he took to be a simple, unanalysable one – in terms of such natural and empirical phenomena as happiness or the survival of the species. It now tends to be used more widely,

with reference to the alleged fallacy of equating values with facts, or of deducing evaluative conclusions from purely factual premisses – editors' note.

2 By a 'formal' cause, Aristotle meant the 'form or pattern' in virtue of which something is the kind of thing it is. By a 'final' cause, he meant the goal of something, 'that for the sake of which' it exists or is done. See his *Physics* 194b23–195a3 – editors' note.

11 Responsibility, ethics and nature

C. A. Hooker

PROLOGUE

The topic of ethics and environmental responsibility is so vast that many volumes might be devoted to it – many have been (see Simmons 1988). Yet here the subject must be confined to a few short pages. Despite my long personal involvement with the issues, my response here has been to focus primarily on orienting the reader to the complex range of possibilities in the topic. I have found a guide of this kind to the conceptual landscape sorely needed, but lacking. So I hope the reader, especially the beginning reader, will find its distinctions a helpful framework. I shall use one specific book, John Passmore's *Man's Responsibility for Nature* (1980), as a reference point for my otherwise rather abstract remarks. This is not the first time I have examined Passmore, or environmental issues;[1] the present remarks are made against the background of these other examinations.

1 RESPONSIBILITY

Remarkably, for a book of Passmore's title and focus, the notion of responsibility is nowhere explicitly analysed therein. Indeed, there is not even an entry for it in the index. Since responsibility is not a simple idea (cf. Vickers 1980), I shall begin there.

The full logical skeleton of a statement of responsibility has the form of a six-place relation, R (A, O, G, X, C, J): A (an agent) takes responsibility for the O (the object entity, system, condition, etc.) to G (the ground of the responsibility) with respect to some aspect, say, X of O under condition C and for the reasons J. A king might order a naval captain to be responsible for circumnavigating the globe, mapping Terra Australis en route and assessing its interest to the throne. Here A is the naval captain; O is the geography of Terra Australis; G is the King; X are those aspects of geography which are of immediate policy interest (perhaps coastline mapping for navigation and coastal prospects for settlement, e.g. water supply, likely arable land); C includes what is accessible to a small sailing vessel supplied with but one or two scientific gentlemen and the scientific instruments they can carry; and the justifying reasons J lie initially in the will of the king and, behind that, whatever justification provides the king with the power which he exercises. So, if we are to acknowledge environmental responsibilities, it will be necessary to identify the six features in each environmental case and show that they go together to form a coherent and defensible claim.

In doing this we need to be aware that the common circumstances of responsibility taking, which I have just illustrated, and which are also the ones emphasized by the *Oxford English Dictionary*, are none the less not the only ones. Consider a husband taking responsibility for his sick wife; to which authority is he responsible? In some cultures it may be a responsibility to a god, assigned by the god, but not in all. (A theistic interpretation of this sort perhaps says more about the projections of authority roles into the religious sphere.) The husband is at the least responsible to himself. And he takes the responsibility voluntarily. Moreover, does not his love say that he is responsible also to his wife – both responsible for, and to, her? We need to be aware of these distinctions when we come to consider claims that we might have responsibility for, or to, the environment.

Finally, note that there are at least two kinds of constraints on the taking/bearing of responsibility. First there is a constraint deriving from the principle that ought implies can: if people ought to meet some requirement (make sound inferences, love their neighbours, etc.) then they must in fact be able to meet that requirement. We don't say that people ought to invent only true theories or stop every conflict they happen upon because in fact they often have no control over these matters. (Note that we would also say that they ought to *try* to do these things; this too is an important distinction in the environmental context.) In the same way we don't hold young children and mentally handicapped people responsible for many actions. So, if someone ought to take responsibility for something then it must be possible in practice for them to do so, and that includes environmental responsibility.

Second, there is a constraint deriving from the justification offered: the conditions under which responsibility is to be exercised must be such that the justifying reasons actually apply. A king cannot legally hold subjects responsible for matters beyond his jurisdiction, such as the actions of another king, or for the salvation of his soul. Similarly with environmental responsibility, we must indicate how the reasons justifying it can actually be met.

Now let us turn to the possibilities for the for, to and why? places in the case of the environmental responsibility relation. (Comments on features and conditions components are implicit in later sections. On constraints, see, for example, Vickers 1980.) It will be complex enough to consider just these possibilities: we may be responsible *for* either nature or just ourselves; *to* either God, nature or just ourselves; and (why?) *because* either God commands it, it is prudentially in our self-interest or it is ethically required. (Here 'God' can stand for any nature-transcendent appeal, and while nature includes ourselves as natural animals it will be simpler to take it here as excluding culture.) The foregoing possibilities make up a space of eighteen alternatives (Figure 11.1).

The Why? axis distinctions need comment. Suppose someone argues that we ought to take responsibility for cleaning up our wastes because the pollution they cause threatens our health, or our monetary wealth. This is a prudential reason for taking on the responsibility; it appeals to our own self-interest, individual or collective. Such appeals make good sense, but they are quite distinct from appealing to ethical or religious reasons to justify environmental responsibility. And why distinguish between appeals to ethics and religion here? First, because there are many ethical atheists whose views are arguably not incoherent and they need to be accommodated. Second, there is an old philosophical argument which concludes that God's commands can't *make* anything ethically right, because we always have the option of doubting God's goodness and the choice of whether to do so is an ethical one, so appeals to ethics are distinct from appeals to religious authority. I shall then distinguish appeals to some transcendental reality (including divine authority), whether or not they proceed through an intermediate ethical practice, from those which are intended to be intrinsically ethical.

We have eighteen possible doctrines to consider, and that's a lot. But before considering them let us note the idiosyncratic character of environmental responsibility; nature is not a person, either little (a

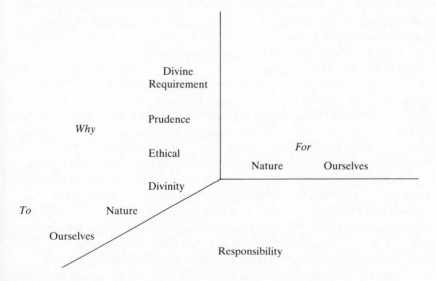

Figure 11.1 Responsibilty: for, to and why?

human being) or big (a god) – let us assume. So we have immediately lost one important part of the normal setting for assigning responsibility. Does it then make sense to speak of taking responsibility for nature, or being responsible to nature, or to appeal to nature as any basis for justifying environmental responsibilities? Moreover, nature doesn't manifest any particular long-term aims – for simplicity I shall assume that is the lesson of Darwinian biology. (For controversy surrounding the a-teleology assumed here, see Hahlweg and Hooker 1988; Lovelock 1979; Nitecki 1989; and their references.) Nor does nature obviously manifest a set of natural values, much less any ethical principles. Though there certainly are *causal* ecological responses when we do anything to or in nature, there is evidently no pattern of responses which has the semantic properties of intentionality. So again being responsible to nature doesn't have the kind of operational interactive sense which being response-ible to a person does. All told, then, one could deny environmental responsibility either on the grounds that you can only be responsible to something sufficiently like a person (so no responsibility to nature), or because you can only be intrinsically responsible for something sufficiently like a person, and hence responsible for a non-person like nature only if some other person (big or little) commands it. Moreover, it is at least not obvious that we *can* be responsible for nature. Humans did not create nature; we remain largely (and dismally) ignorant of nature's constitution and dynamics; we remain powerless to alter many (but not all) of nature's basic processes, and the attempt to preserve nature may well disturb ecological processes anyway (Canada was an ice sheet only 15,000 years ago). So before leaping into heartfelt talk about environmental responsibilities we should think carefully through what is intended by the idea.

Passmore, for example, thinks that our primary, essentially our only, responsibilities are to humans and directly only for humans, not directly to or for nature at all. Nature only enters as the indirect object of the responsibilities. Thus Passmore's position lies among the For = Ourselves, To = Ourselves positions on Figure 11.1. Many of Passmore's main arguments for pollution control measures, population control and conservation are based on our own self-interest, on the damage we will inflict on ourselves unless we take measures to protect the environment. This is really taking responsibility for our own actions, not for the environment; nature simply continues to 'do its own thing'. This position lies at the Why? = Prudence location and is in fact the commonest position. It provides the minimal basis from which we might try to justify taking some (indirect) responsibility for nature. And in the justifying reasons we can cite all of the widespread environmental damage and ecological destruction with which we currently threaten the biosphere and, through that, ourselves and our children. This case for such responsibilities has been made often and well. In this chapter I am ultimately interested in what ethical basis there might be for environmental responsibility. But before coming to that it is well to note that there is a more ancient religious alternative which, while it agrees that nature is only an indirect player, is interestingly different from Passmore's prudential position. And Passmore himself describes it well. Let us look briefly at it.

2 RELIGIOUS RESPONSIBILITY FOR NATURE: THE JUDAEO-CHRISTIAN TRADITION

The Judaeo-Christian tradition has most often been blamed for promoting a careless and rapacious attitude to the environment. 'Go forth and multiply and subdue the earth and have dominion over it.' Here we have a projection of empire mentality on to our relation with nature ('dominion'), humans (man!) as conqueror of nature ('subdue'), hence nature as useful possession, to be overwhelmed and displaced ('multiply'). Humans here are separate from nature, and superior to it. In this tradition we have responsibility neither to nature, nor for nature. We are responsible to God, for ourselves.

However, the 'man-as-oriental-despot' theme is not the only one in the Judaeo-Christian tradition. In particular, Passmore isolates a 'man-as-steward' theme, which is conservationist in attitude, and a 'man-as-creator' theme which is more boldly interventionist in attitude and looks to humanity as a perfector of nature. Both themes, the former especially, recognize nature as existing to glorify God, nor merely to serve humanity. Both themes lead to a view of nature as valuable in itself as well as valuable for humanity.

According to the 'man-as-steward' tradition we have responsibility to God, for nature in its own right, as well as for ourselves. The ultimate basis for our bearing this responsibility may either be taken as God's command or because of the intrinsic value nature has been given as a glorifier of God. In either case this position is at the For = Nature, To = Transcendent, Why? = Transcendent location. We could read the 'man-as-creator' tradition in the same way. But we might extend it to the view that we had a responsibility to nature itself to perfect it, because of the kind of intrinsic value it had. In that case it would be found at the For = Nature, To = Nature, Why? = Ethical location.

This last position is the first time anything distinctively ethical has emerged by way of a basis for environmental responsibility. But before we examine ethically based responsibility, let us briefly look at the commonest position, that based on prudential considerations.

3 DERIVED OR PRUDENTIAL RESPONSIBILITY FOR NATURE

Not every environmental action is based on intrinsic concern for the environment. To the contrary, most are not. Perhaps the commonest basis for environmental action is a belief that inaction will rebound to harm humans, and in particular the actors themselves. Pollution produces poisoned drinking water and food, ruins sports fisheries, damages recreational forests and so on. Tropical forest destruction removes opportunities for new medicines and improved food crops, increases the pace of greenhouse warming and so on. These are all very good reasons for considering action to protect the environment, perhaps the only effective reasons for doing so. In any event they play a leading role in current environmental concern; they are leading us to develop a very important set of environmental mitigation policies.

But they do not stem from, nor lead to, any intrinsic concern for the environment as such, whether ethically based or otherwise. Instead, these considerations are all prudentially based; it is in our own self-interest, either as individuals or as a community or species, to take these actions. Were this self-interest absent – because, say, destruction of some ecologies produced no effects harmful to us – then there would also be no prudential basis for taking any care of nature. Of course one could still differentiate among ways of destroying nature; the conversion of countryside into ugly concrete jungles might be considered 'bad', but its conversion into English farm meadows or elegant villages might well be considered 'good'. In Figure 11.1 these positions occupy the location For = Ourselves, To = Ourselves, Why? = Prudence.

As noted earlier, Passmore's main arguments for pollution control measures, population control and conservation all fall into this category. But prudent self-interest provides powerful reasons for protecting the environment (in certain respects) and Passmore's discussion of the issues is an excellent place to begin. None the less, there is here no ethical obligation to the environment as such; it still figures here as something of instrumental value only. Indeed, it is striking that environmental sustainability does not figure among the valuable characteristics of western civilization which Passmore defends as central to what should be passed on to the next generation. (He concentrates on diversity of culture, humanitarian practice and degree of liberty – see his pp. 182–6.) Of course even sustainability is primarily, and perhaps wholly, prudential.

But Passmore also offers an ethical basis for assuming general environmental responsibilities which derive from obligations to our future selves or our future children. In the spirit of exploring the bases for environmental action I want to turn now, therefore, to the notion of environmental ethics.

4 ENVIRONMENTAL RESPONSIBILITY AND ENVIRONMENTAL ETHICS

The essence of Passmore's ethical argument is that, out of justice and (more powerfully) love, we should hand on to the next generation a world whose condition makes it possible for them to fulfil themselves as human beings, according as we best understand this latter term (and including, centrally, their capacity to choose among at least the ways of being human that we find it reasonable to cherish). But though these are clearly ethical responsibilities (in the wide sense of ethics – see section 5), they are not intrinsically environmental ones. The environment is nowhere given any intrinsic role. As it happens the actions called for benefit the natural environment in a wide range of cases. But had they not, perhaps because it was considered more just or beneficial to future generations to hand on an artificial, manicured environment free of the threats and uncertainties of wilderness, then the basis for our responsibility to the future would none the less remain intact. Thus Passmore's position here still falls into the category For = Ourselves, To = Ourselves, since it is directly concerned only with human/human responsibilities. It is time then to investigate the notion of an environmental ethics more widely.

Though Figure 11.1 assigns a simple, single-line entry to ethics, the notion of ethics is itself complex. First I want to say some things about the scope and complexity of ethics. Then I want to characterize the range of environmental ethical positions or systems which are currently under discussion. Then I shall say something briefly about the distinctive issues in environmental ethics.

First let us distinguish all those aspects of life which have these two features: (i) they arise from what we deeply value and (ii) they are intended to transcend merely individual points of view, interests and imperfections (such as bias, carelessness, egocentrism, anthropomorphism). Plato held that the three deepest values were goodness, truth and beauty and that their pursuit in a way which transcended our egocentric imperfections was the mark of the life of reasons which issued, respectively, in ethics, sciences and aesthetics. Ethics then I shall take to be the pursuit of goodness in a way which transcends individual interests and human imperfections. What then might go into ethics, thus understood? A number of diverse components, as given in Figure 11.2.

Briefly, I shall take it that the central idea of justice is fairness, as some kind of equality of treatment. Then equality divides between process and product, there can be equal *opportunity* for X (process) and there can be equal X *outcomes* (product). The economic market aims at equal opportunity for wealth; communism aims more at equal wealth. A similar contrast applies to access to the law versus legal judgements. As these examples suggest, these two forms of justice tend to conflict with one another, though both are valuable. Rights I shall take to refer to those conditions and actions to which their holders ought to have access if they so choose, such as the right to be free of externally inflicted pain. Rights may be assigned both to individuals and to groups (e.g. nations). Assignments of rights often mutually conflict, as when a right to free

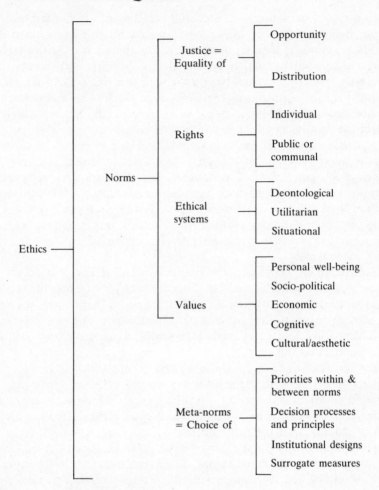

Figure 11.2 Components of ethics

assembly exercised by several removes someone else's right to peace or conflicts with a national right to restrict entry. See the relevant entries in *The Encyclopaedia of Philosophy* (1967).

Briefly too, a deontological ethical system is based on obligatory rules, such as 'Do not murder'; a utilitarian ethical system considers the good action to be that which maximizes human welfare (however measured); and a situational ethical system avoids general rules entirely but aims at promoting the flourishing of each individual in each specific setting. Of course the welfare-maximizing action may well be one forbidden by the deontological rules (e.g. murdering a violent dictator) and in conflict with promoting the flourishing of the individuals concerned (both the dictator and the murderer suffer), so that these approaches diverge. Values, though actually complex, I shall here take to be clear enough. But they too can certainly conflict among themselves (e.g. freedom of speech versus public peace). Finally the meta- category concerns the normative judgements which necessarily underlie our choices among and within these categories, either in assigning precedence among them for the purpose of conflict resolution (see p. 161), or when designing the institutions and technologies which shape the structure of conflicts as well as the resolution processes (see section 6).[2]

The important point to grasp about these components to ethics is their potential diversity. As noted above, within each component there are conflicting attractive options. And components also conflict among themselves across Figure 11.2. Thus the right to free assembly may lead to an ecology being trampled (rights versus values); maximizing aggregate (or even mean) welfare might leave some very poor (justice versus utilitarian ethics); valuing loving leads beyond what justice requires (Passmore 1980); and so on. We should then abandon the idea that ethics consists in searching for a single correct response to a given situation which would somehow satisfy simultaneously all of the attractive values and principles. To the contrary, it will be of the essence of responsible ethical judgement (at least in this finite life) to know how to balance values and principles against one another intelligently.

If this complexity to ethics were not sufficient, there is an equal complexity to the way one approaches the terms within which ethics is developed. The essence of the issue here hinges around whether ethics is primarily formulated in terms of individuals or whether it is primarily formulated in terms of such whole complex systems as societies and ecologies. Of course, there is no sharp separation between these two extremes; they are connected by a continuum of positions. To set up the relevant alternatives, distinguish between the classes of individuals or systems which have features making them worthy of ethical consideration and the class of individuals or systems on which the ethical principles focus. Then we find the array of positions sketched in Figure 11.3.

What features make something worthy of ethical consideration? At

Ethical-worth features

		Egocentric	Anthropocentric	Ecocentric
Focus of ethics	Individual	Shallow ecology (traditional liberal)		Animal rights +
	Communal	Pastoralist Marxist		Deep ecology

Figure 11.3 Categories of environmental ethics

one extreme one may hold the view that it is only those features which contribute to something's being an individual agent, like a human being, that are relevant for ethical consideration.[3] At the opposite extreme are those who hold that the features which give primary ethical value to something are whole system features such as the justice of a society or the complexity of an ecosystem. They would count individual features of less importance in ethical consideration than these larger characteristics. Similarly, it is possible to hold ethical principles focused wholly on obligations or responsibilities to individuals and expressing responsibilities for individuals. On the other hand, it is possible to insist that ethical principles should focus primarily on obligations and responsibilities to, and for, systems as a whole, whether societies or ecosystems.

The traditional ethical stance focuses on individuals ('*thou* shalt not kill (*an individual*)'; maximize welfare across *individuals*) and counts only humans as having specifically ethical value. The environment is of only indirect interest. Hence the term 'shallow' applied by some. The animal rights movement wishes to extend ethical consideration to other creatures, thus enlarging our circle of ethical concern, but leaving the focus of ethical principles on the individual. (Thus, continue to maximize welfare across individuals, but now include the welfare of individual animals as well.) By contrast, the Marxist tradition emphasizes the priority and importance of the community over the individual. So, while Marxism has no special regard for the environment – humans remain central – it emphasizes obligations to the community rather than to individuals. Similarly, the Christian stewardship and creator traditions belong to a communalist strand of Christian thought and underwrite a shift towards primary communal ethical obligations, with humans still the focus of ethical concern. These and like traditions I have labelled 'pastoralist'. Finally, there are those who wish to combine the enlargening of the circle of ethical concern with a communalist orientation to obligations; they are self-labelled the deep ecologists. I shall have more to say about them, and about the animal rights position, in section 6.

So far I have tried to provide an outline of alternative approaches to environmental ethics. But it is easy to confuse the rather general level at which I have been speaking with the several other levels of ethical claims, and with practice. These are outlined in the following list. I shall then provide a very brief commentary on them.

- *Meta-ethics*: theories of the nature of ethics itself, e.g. that ethical principles are divine commands, or biological necessities, or the outcomes of rational self-interest, or strong intuitive truths, or merely express subjective emotions;
- *Ethics*: theories of obligation or responsibility or rightness, etc., as indicated above;

- *Applied ethics*: combinations of ethical principles with other principles, e.g. economic, legal or political principles, and with factual information, to develop practical principles for some specific situation or kind of situation, e.g. hospital treatment of premature babies, logging rights *vis-à-vis* watershed conservation;
- *Morality*: the actual practices of a family, community or culture which manifest, and support, some ethic, e.g. recycling, punishing lying, praising generosity.

In the longer run it will matter a great deal what meta-ethics you adopt. If you believe, say, that ethical principles describe those ways of behaving that evolution favours then you will be led to consider some explicit connection between ethics and environment. (But beware that it might not be environmentally protective! One version has it that it is the ruthlessly selfish that survive best.) If, by contrast, you view ethical principles as divine commands then, as we have seen, there is no necessary connection to the environment; it all depends on the religion in question. And if you view ethics as those social principles that best promote individual interests, as the general outcomes of a rational bargaining process about how we want to live, then there will be a kind of collapsing of ethical environmental concern into prudential concern for how our environment affects us. (At least this will be so unless a community includes love of nature for its own sake among its publicly supported values.) As you can see, the effects of meta-ethical views are complex and important; but there is no space to pursue these issues further here.

Ethics we have already seen to contain a lot of complex, conflicting components. The complexity increases much more again when we come to applied environmental ethics because of all the new factors to be combined with ethical principles. Partly this is because other principles may conflict with ethical ones. The law may prevent you applying sanctions without evidence of wrongdoing, but science may inform you that by the time environmental wrongdoing has shown up it will be too late to save the ecosystem in question. And of course preserving wilderness will often not be economically profitable, neither to any private enterprise nor to government. (But if ethics is to express collective rational self-interest, it has somehow to be profitable to something – here you see the further impact of meta-ethics.) An environmental ethic can as easily also conflict with social or political principles.

In part, the complexity of applied environmental (or any other applied) ethics arises from the factual complexity of real-life situations. It is strictly impossible, for instance, to create general standards for safe discharge of even a single industrial pollutant into waterways; what is acceptable to various non-food agricultures, to various food agricultures, to other individual industries, for human drinking and for ecological conservation will in general all vary, and vary also with the natural dissolved contents of each river and lake. In practice the best we can do, to make this complexity manageable, is to combine general guidelines with particular regulations as each case demands. Again, this is all the commentary on applied environmental ethics which space permits. I turn next to practice.

Without a suitable practice of a set of habits, attitudes, feelings, institutionalized roles and so on which make up a coherent 'lifestyle' no ethics would be expressed in practice, or reinforced. I have introduced the term *morality* to designate these practices, which are distinct from the theoretical content of ethics as such. Morality ranges from recycling paper to public disapproval of lying, from families encouraging loving feelings to the imprisoning of criminals. A crucial part of environmental responsibility is the development of an environmental morality. This will include recycling resources; educational fostering of knowledge of, and respect for, nature; engaging joyfully in benign and sustainable lifestyles; developing and enforcing an environmental legal code; and all the host of activities that are involved in putting an environmental applied ethics into practice. But I shall also say no more about environmental morality here.

5   WHAT A THEORY OF ETHICAL ENVIRONMENTAL RESPONSIBILITY INVOLVES

To have a *coherent* position regarding ethically based environmental responsibility requires selecting one of the alternatives presented in Figure 11.1 and one of those presented in Figure 11.3 together with a consistent set of instances from all the categories of Figure 11.2 and a similar selection from all categories on pp. 156–7, such that these four selections mutually cohere to form a workable body of theoretical principle and moral practice. How many alternative positions are there? Very many, as many as there are coherent combinations of doctrines from these four sets. Even so, these doctrines still exclude all of the non-ethically based doctrines of environmental responsibility, e.g. the purely prudentially based ones. So there are still more alternatives available.

I have tried to sketch the variety and complexity of theories of environmental responsibility because in the literature only particular fragments are usually discussed, e.g. some principle of justice *vis-à-vis* future generations. While this is valuable, one is left in the dark as to precisely where any one fragment might fit in an overall position. To indicate how development of a theory of ethical environmental responsibility might be approached, I shall conclude this chapter with some critical remarks on what I take to be the major ethical alternative to prudential considerations.

6 LOVE, CONFLICT, DESIGN

Passmore argues, recall, that love for our children can bind us to do for them what justice, and certainly not self-interest, could not bind us to do. This includes bequeathing them a healthy environment. I should like to accept and generalize that insight. I suggest the following propositions for your assent: Love is the ultimate ground for all ethical considerations. Love moves us to actions beyond those which any impartial rules enjoin. Indeed, to become loving is one of the fundamental goals of life. Love must provide the ultimate ground for environmental responsibility.

But what now is implied by loving for environmental responsibility? In response I first draw an important distinction between two kinds of love, eros or ego need-love and agape or gift-love. Both are constructive, but eros is the narrower. Eros is the expression of the creaturely self, the need both for recognition, acceptance and pleasure and to give these to others reciprocally to express self-integrity. Agape is the freely cherishing of another life for its own sake. For humans, the development of eros is essential to the formation of a healthy personality, and to the capacity for agape, but it is agape which crowns personal or spiritual, and so moral, development.[4] Environmental responsibilities based on prudential considerations stem ultimately from eros. (So might that based on religion if it is driven primarily by fear of punishment or desire of reward.) Such considerations are, I have noted, basic (section 4); but here I shall focus on what might follow from agape.

There is certainly the implication which Passmore draws from love of humans, that we are love-bound to hand on to our children an environment which permits them the opportunity to flourish as human beings. But, as Passmore makes clear, spelling out the detailed implications of this is complex; no simple environmental policy follows. Should we aim to hand on England's green shores, long ago transformed into an artificial human-made environment, or Canada's wildernesses? And granted that our children may differ from us over what constitutes human flourishing, which features should our bequest be sure to display and which are dispensable? It is important to follow through these complex issues, but the larger point is clear: to love the world already distinguishes the lover and the loved. The very consciousness which makes possible the life-project of learning to love also distinguishes us from our environment, and raises the question of whether our goals coincide with any particular process in nature. Note that the same conclusion issues from our capacities as creators, since what is created is our making and not simply nature's product. (This is a conclusion of some importance. Every species transforms its habitat; we can do so on a grand scale. And evolution has given rise to just this capacity. This gives a special complication to understanding what is natural, and a special importance to design; see p. 163.) So then, we should aim to love other humans (and be creative), yes, but does the rest of nature figure any way but indirectly in the picture?

There is an important argument for a direct role for nature in our loving, the expanding circle argument, put powerfully by Singer (1981). In the past (and, alas, still today), the argument runs, it was common not to include all humans in the scope of ethical obligations. Rather, obligations typically extended only to one's own group (extended family, tribe, etc.). Others were treated callously (slavery, ruthless warfare, etc.) and even denied human status (cf. Nazi attitudes to Jewish people). We are slowly coming to learn, however, that the circle of ethical care should be expanded to include all humans, irrespective of race, religion, sex, etc., and achieving this, in entrenched legal or political codes, for example, presents significant ethical progress. Well then, the argument continues, we should now come to see that the circle needs to be expanded still further, this time to encompass other creatures that can feel pain and pleasure. Making them equally part of our ethical concern also represents ethical progress for us. From such arguments grow the animal rights positions which Singer and others have championed (Singer 1975; Regan and Singer 1976).

Notice that this is not a proposal that animals and other creatures should be treated as moral agents; recalling the two distinct components of Figure 11.3, the proposal is that other creatures should figure among those to be given moral consideration. Which features should entitle a species to moral consideration, and to how much consideration? Here there is controversy; for some the key feature is the capacity to have goals and intentions, for others the capacity to feel pain and pleasure, and for others (embracing still more) it is the having of needs at all. Again it is worthwhile to follow these complexities through, but however they come out, it seems to me that there is something right about this idea. For any human knowingly to allow another creature to suffer pain unnecessarily, for example, surely diminishes them spiritually. They are shown to be less than fully aware, and not only of others, but also of their own marvellous conscious selves. To understand gift-love we must be aware of (and, it turns out, have gift-love for) ourselves. But once gift-love is grasped at all it is hard to confine it to a narrowly defined group of creatures.

None the less, this approach has been criticized by 'deep ecology' and the criticism is worth considering for the way it helps to focus a second dimension to loving. The essence of the so-called deep ecology position grows out of two claims: (i) our spiritual development calls us to become one with the world (the cosmos) as a whole in love and we are, therefore, ethically responsible for treating the whole with due respect and ethical consideration; and (ii) the important features for ethical consideration are whole system features, overall properties of ecologies and societies, rather than of the individuals which make up those systems. Their view of responsibility might best be captured by To = Nature, For = Nature, Why? = Ethical (Figure 11.1). In the literature there is diversity of proposed norms within deep ecology (cf. Figure 11.2), and a general tension in doctrine (i) between, roughly, a Buddhist and a Christian reading. However, discussion is often of more concrete principles at the level of applied ethics or morality – for instance, Leopold's 'land ethic' (Leopold 1949). (See e.g. Devall and Sessions 1985; Evernden 1985; Routley and Routley 1979, 1980; Tobias 1985.)

In common with the expanding circle/animal rights position, deep ecologists charge traditional ethical positions (Figure 11.3, left-hand side) with anthropocentrism. But because of assertion (ii) above they also charge the animal rights and allied movements (Figure 11.3, right top) with inappropriate individualism. They argue that these defects appear all through the normative categories (see pp. 156–7) in the common run of environmental literature and argue for eliminating them everywhere.

It is the call to oneness, claim (i), which really drives claim (ii) and the whole position. And here a further dimension to gift-love emerges, the drive towards union of the lover and loved. Ultimately the lover is 'lost' in the beloved; identity is submerged in – better: diffused through – the beloved. This happens in the small when two lovers create, over time, two new selves each now a complementary part of a single, richer union. (It is what gift-loving is about for finite, imperfect creatures, and it does take time, and ego pain – insights our culture seems in danger of losing.) It happens in the large when the enlightened achieve mystical union with the godhead, the Good, the One (cf., for example, Huxley 1958; Wilbur 1981; and references). It is this drive which is, I suggest, reflected in the eco-centredness of deep ecology. The deep ecologists' basic argument is that we cannot truly attain our moral maturity without acknowledging the claims of gift-love towards all life. Schweitzer, who gave up brilliant European careers in medicine and music to run a bush hospital in Africa, regarded the whole of life as sacred; the ethical person, he says, will not even tear a leaf from a tree unnecessarily. And this too has something right about it.

But it also needs to be tempered with another profound aspect of life for creatures in this finite, imperfect world: the ineliminable presence of tension. Schweitzer, for example, chose to cure humans rather than animals, and he chose not to further medical research science. This is typical of the tensions in ethical choices. (See Schweitzer 1949, reprinted in Regan and Singer 1976; cf. comment by Singer in Goodpaster and Sayre 1979.) Here I turn from the theme of love to that of conflict.

Some conflicts are straightforwardly resolvable through compromise, for example in economic bargaining. Some conflicts are resolvable by transcending the terms in which the conflict originally arose – for example, when lovers agree to alter their values so that a conflict between them no longer arises. And some conflicts reflect what I call tensions, irresolvable conflicts between values or goals, each of which is in itself good but each of which interferes with achieving the others, as in the case of Schweitzer. (On the distinction between these two kinds of problems, and both from tensions, see Schumacher 1977; cf. Schumacher 1975. On some tensions, see Hooker 1987: chapters 7, 8.)

Before turning to tensions, I note that the strategy of transcending the conflict through co-ordinated value shift is one which the prudential approach tends to ignore. None the less it is fundamental to the conduct of life; it is, in my view, what lies at the foundation of all social institutions. And it is, therefore, central to any realistic theory of environmental responsibility. Briefly, social conflicts are resolved by providing institutionalized roles, with accompanying rewards and punishments, such that our resulting individual self-interests coincide (roughly) with the communally optimal condition. In this manner, and essentially only in this manner, is the important class of games called prisoner's dilemma games resolved. All environmental problems present themselves to us as prisoner's dilemma games. Hence the importance of institutional design to environmental policy.[5]

Every aspect of life is replete with tensions. Ethics, for example, is replete with tensions, as the discussion of Figure 11.3 revealed. All of these tensions are manifest in the domain of environmental responsibility. Thus granting rights to animals of kind A will on many occasions conflict (i) with granting rights to ecologies (whenever specific animals would disrupt an ecology, etc.), (ii) with human rights (e.g. of free assembly in a feeding area for As), (iii) with providing justice for both humans and animals of other kinds (e.g. when some group of them must be denied equal treatment in some respect in order to respect A's rights), (iv) with maximizing human welfare and/or the welfare of other animals, and so on.

The same applies to any other ethical claim here, such as granting rights to, or moral value to, ecologies. (And what ecological features shall we value? Complexity, stability, speciation have all been proposed, but each may conflict with the others.) Some of the more specific tensions which apply to love-based environmental responsibility are those between granting rights to individuals and cherishing systems as a whole; promoting benign artificial environments (gardens, etc.) and promoting wilderness; maximizing future human environmental options and maximizing present environmental quality for humans; promoting conservation of resources, preservation of environments and human wealth (using Passmore's terms); preserving species and preserving the evolutionary process (which eliminates, not just individuals, but species); and so on. Behind all of these loom larger tensions, principally that between the desire to be a unique self and the desire to be one with others and with the world.

These conflicts will not go away. The best we mortals can do is to seek reasonable momentary compromises that resolve them for the specific circumstances to hand, always ready to adapt them to new situations. This unsatisfying necessity does not invalidate environmental ethics, but a convincing theory of ethical environmental responsibility will clearly recognize them and pursue their current best resolution. One of my criticisms of deep ecology is that it tends to have too simplistic a conception of our ethical circumstances.

The resolution of tensions also turns us to the question of design. I hold that, like our compromisable and transcendable conflicts, we also resolve our tensions through designs, principally moral, institutional and technological designs. So a theory of design becomes central to a theory of applied environmental ethics. We need good institutional designs, for instance, involving our systems of economics, law, administration, etc., to resolve constructively the economic prisoner's dilemma games which lead to environmental disruption. We need good designs for democratic process to allow us to work responsibly through the compromises necessary to resolve our ethical tensions over environmental care, like the compromises with individual freedom about which Passmore is so concerned. And so on. Beyond all these we need good cultural designs, involving local community structure, educational process, etc., to promote the basic processes through which people come to be more loving. Otherwise the other institutional designs will become the empty, merely procedural shell they are in danger of becoming in our societies. (Here I can only indicate the profundity of the cultural design issues by pointing towards some diverse literature for readers to follow up; see e.g. Chance 1988; Eisler 1987; Vickers 1970, 1980, 1983.) It was the absence of a serious theory of design which was my severest criticism of Passmore earlier (Hooker 1982). Now I should extend that criticism to many other accounts of environmental responsibility, in particular to deep ecology.

7 CONCLUSION

I have given no comfort to those seeking relief from complexity and compromise in simple slogans and solutions. I will have disappointed some by subordinating the statement of my own views to the provision of analysis and disappointed others by not advocating a specific line of action. (To catch my own views and actions you must read 'between the lines', and consult the references.) But I hope to have done my readers the greater service, in a confused and complex area, of providing a wider framework within which they can begin to think through the issues for themselves.

NOTES

The author is grateful to Professor R. E. Butts and the Department of Philosophy, University of Western Ontario, for provision of facilities during preparation of this chapter, and to professors Barry Hoffmaster and Michael Yeo for helpful criticism of its ideas.

1 On Passmore, see Hooker (1982) and for my other writings see the References. Material in sections 1–5 of this chapter is drawn from Hooker (1990a).
2 These latter include the choice of decision methodologies, e.g. cost–benefit analysis, as well as the measurable surrogates, e.g. for environmental values, which those methods demand. On cost–benefit analysis, see, e.g., Hooker (1984b).
3 There is a finer distinction here between those who would exclude any non-human creatures and those that would include all creatures which are relevantly like human persons, and there is a converse issue of how many of these features a human individual can lack while continuing to be worthy of full ethical consideration (e.g. how mentally deficient); but I shall set these fine-nesses of distinction aside here for the moment. Frankena's discussion in Goodpaster and Sayre (1979) contains these and other helpful distinctions.
4 On the agape/eros distinction, see Nygren (1953). Agape is the near equivalent of joy in the eastern traditions. On personal development from the agape perspective, see, from amongst a large literature, Wilbur (1981) and Erikson (1963).
5 See e.g. Hooker (1983) and Hooker and Van Hulst (1979, 1980). The notion of design here includes technological designs, with technologies modelled as amplifiers in institutionalized contexts/systems. See e.g. Hooker (1984a, 1987). For application to energy policy design, see e.g. Hooker et al. (1980) and Hooker (1989); and to health care, see Hooker and Van Hulst (1979) and Hooker (1990b). But these considerations must be complemented by those of value transformation through institutionalized moral practices; see e.g. Vickers (1970, 1980, 1983).

12 The idea of environment

David E. Cooper

I

Recent books on environmental issues usually inform readers of the emergence, during the last twenty years, of a 'new' environmental ethic, sometimes going by such names as 'deep ecology' or 'ecophilosophy'. This 'new' ethic is prominent, for example, in current discussions of 'sustainable development': a familiar line being that, if this is to be more than the attempt to spin out industrial growth for a few more years, it must be converted by a moral perspective into 'reverential' or 'integral authentic' development.

Why has it seemed important to promote a 'new' environmental ethic? The popular answer is that, without one, we shall continue a plunge towards catastrophe. A 'shallow' policy of prudent economic development is too close in spirit to the exploitative urge which sent us flying in the first place to break the fall for long. Left by itself, this answer has a disingenuous ring, for it proposes a 'new' ethic, not as moral truth, but as a convenient myth, a holy lie, one of Nietzsche's 'falsehoods without which men cannot live'. But as a term like 'ecophilosophy' might suggest, there is a further answer. A 'new' ethic is required not only on pragmatic grounds, but by a true appreciation of the place of human beings in the world or 'ecosphere'. 'Sustainable development' will only be 'authentic' when it is faithful to this appreciation.

I am not alone in being depressed by the monotonous character of much of the literature, especially at the more popular end of the spectrum, which declaims the 'new' ethic. Like Gore Vidal's, the eye can easily 'glaze over' (even if, unlike his, it does not detect a conspiracy to substitute a new enemy, man himself, for old ones like communism) (1989: 29). It is even possible to sympathize with one critic's desire to 'see someone on TV passionately demanding that the Amazonion wilderness is a blot that should be erased from the Earth' – not because one agrees with the demand, but because it might shake the 'self-righteousness and blandness' of the currently monochrome consensus.[1]

The following passage may smack a little of parody, but it would not be an unrepresentative pot-pourri of the 'new' ethical themes to be found expressed in countless books, articles and broadcasts:

> Moral progress has consisted in widening the moral constituency, with the gradual recognition of the rights of foreigners, women, people of different races, and so on. Such progress is stunted, however, if it remains anthropocentric. Indeed, it cannot rest until the moral status not only of flora and fauna, but of whole ecosystems, is recognized. A new eco-ethic is, therefore, an extension of justice. Such an ethic will promote attitudes of reverence, awe and respect towards living things and systems, the whole environment in fact. For these have their own sanctity and intrinsic value. Underlying these attitudes will be the appreciation that our world is a seamless whole, an eco-sphere, Gaia, in which every being, including man, has its integral place. The ethic will be a naturalistic one, since it is gleaned from the observation – no longer distorted by anthropocentric science and philosophy – of natural systems. Only such an ethic will have the force to reverse the process of environmental destruction.[2]

The monotonous repetition of these themes would hardly matter, of course, if they embodied clear and evident truths. But they do not, and my sketch serves to hint at some of the defects of much 'new' environmental writing.

To begin with, there is the incantation of a pious vocabulary – 'reverence', 'awe', 'sacred' and so on – which can soon begin to sound hollow. I do not revere or hold sacred the Amazonian rain forest, not because I am irreverent or profane, but because, never having been remotely engaged with that forest, such a vocabulary is inappropriate coming from my lips. For me to speak that way devalues the same vocabulary when spoken by those whose home the forest is. Again, if Andrew Marvell's verdict on mountains, 'ill-designed excrescences', is barely intelligible to a later age with different aesthetic and religious predilections, the blanket demand to look on everything in nature with awe is no less perverse; for it would remove all point to describing any particular natural phenomenon as awesome.

Second, there is the all too rarely remedied vagueness of the underlying 'holistic' theme of everything in nature, ourselves included, being integral elements in a seamless whole. On some interpretations, at least, this is clearly trivial or implausible. It is without interest and moral implication that a falling tree in Australia has *some* connection with the birth rate of flies in Alaska; and it is false if it is meant that such connections must be detectable, let alone significant. This is not to deny the importance of ecological insights into feedback and interaction; but this resides in directing attention to the study of particular systems, not in a sweeping claim about the relatedness of everything with everything. One wonders, too, at the insistence on man being 'part of nature', given an accompanying insistence on uniquely human capacities, such as moral reflection, which one might think place him 'outside' nature.

Finally, the 'new' ethic embraces diverse themes among which there are real, yet generally unremarked, tensions – at both practical and theoretical levels. There is, for example, no obvious compatibility within a single ethic of the 'deep ecologist's' concerns both for conservation and for economic justice for Third World nations – not if justice here means increased wealth for these nations and if wealth is a prime factor in over-consumption.[3] At a theoretical level, one suspects tension between the attitude of 'reverence' we are urged to accord nature and the 'holistic' theme of man as just one 'part of nature'. For, as some 'new' writers themselves stress, that attitude might only be appropriate towards what Rudolf Otto called 'the wholly other'. Worship of a nature that includes ourselves might betray that hubris of which 'new' ethicists typically complain.

II

There is no pleasure in raising these criticisms, since the heart of the 'new' environmentalists is surely in the right place. Indeed, an aim of the thesis I shall develop is to salvage the element of truth in their approach. I want, first, to identify a main source of the hollowness, vagueness and tensions of the 'new' ethic. This will enable me to deepen the criticisms just raised and to suggest the outlines of a more promising approach.

The source of the problems, I think, is the distended notion of environment which imbues the 'new' ethic: one which, ironically, embodies just that scientific perspective against whose dominance these writers rail. Put crudely, their notion of environment is of something much too *big*. The environment for which we are supposed to feel reverence is nothing less than nature itself, and it is the whole natural order of which we are urged to regard ourselves as integral parts. Each person's environmental concern is supposed to extend everywhere, 'from the street corner to the stratosphere' as a currently popular adage has it. The expression 'the global environment' makes this largeness of scope explicit; but even when the adjective is omitted, the definite article and singular noun indicate that there is just one big environment – the biosphere, the order of things. From now on I refer to the environment so conceived with suitably large letters – 'The Environment'.

The first thing to establish is that The Environment is very different from an earlier and once prevalent concept. This is quickly seen when one reflects that wildernesses, very much parts of The Environment, used to be precisely what lay outside people's environments. An environment, etymologically, is what *surrounds*; but 'surround' was never understood here in a geographical, geometrical sense. In *that* sense, I am surrounded by everything located in a circle drawn around me: by everything there is, in fact, if the radii are long enough. The relevant sense, rather, was the one suggested by various foreign equivalents to 'environment', ones which have, of course, found their way into English – such as the French *milieu* or the Spanish *ambiente*. Birmingham and Malta are both surrounded, geometrically, by the sea, but only in the latter case is the sea a milieu, an environment. Fish which spend their lives patrolling a few square yards of coral are in the ocean, but only a tiny fraction of the ocean belongs to their ambience, their environment.

These examples mislead if they suggest that the crucial feature of an environment, as once understood, is proximity, short radii. The beech at the end of my garden belongs to my environment, despite being further away than the rubble beneath the floor where I stand, which does not. A badger in a set by a motorway has the trees on his side in his environment, even though some of these trees are further away than the ones just over the road. City dwellers from different sides of the tracks may share an environment less than hill farmers living miles apart.

The older concept is not, therefore, that of the *local* environment: not if this is measured in yards or miles. We get a bit closer to that concept if we think of it as that which has considerable causal impact on a creature, irrespective of mileage. This, one might think, is why the trees on the other side of the motorway are outside the badger's environment. But this cannot be the whole story. The sun has massive causal influence on all life, but (with a qualification to come) it was not considered part of a creature's environment. Conversely, the scents in my garden are an aspect of my environment, but unless I am allergic to them their significant impact on me is not *causal*.

It is not, of course, chance that an environment is associated with what is geographically close to, and has causal impact upon, a creature. But these do not belong to the essence of the older concept, the associations being due, rather, to certain contingent facts about creatures, such as the range of the sensory organs. I shall explain this later.

What, then, is the once prevalent idea of an environment? It is not, unfortunately, a sharply delineated one, and despite (or because of) its pervasive role in people's lives, it has rarely been articulated. Let us take our lead from those terms – 'milieu', 'ambience', 'neighbourhood' even – which, until recently at least, were close relatives of 'environment'. These terms denote what a creature knows its way about. A creature is not in its milieu or ambience like a currant in a bun, but more in the way a pupil is in a particular school. The pupil knows how to get from A to B, how to address people, whom to avoid, what to wear and which feelings to express or disguise. He or she is, one might say, 'at home' there. Likewise, the badger is not in its set in the way a clod of earth is; the roof of its tunnel is the roof of its home, with all this implies for its relationship to the surroundings. The badger, too, knows its way around, how to treat other badgers, which animals to avoid and perhaps how to dissemble.

An environment as milieu is not something a creature is merely *in*, but something it *has*. This is why it can find itself without one, as when I am parachuted into the Sahara or a badger removed to a laboratory. Neither of us then knows the way about; nothing is familiar or has anything of home for us. A creature without an environment would, of course, be an impossibility if the only sense of the term were that of The Environment or geographical bits of it. Each creature must, after all, be somewhere and not nowhere. But this only reinforces how different the notion of The Environment is from an older conception.

On that conception, an environment is what a creature knows – and knows in a certain way. Armed with a map, I can find my way around a strange terrain, and a zoologist might in his way know more about the set than the badger does. But these are not the kinds of knowledge which make a terrain or an underground tunnel part of an environment. The relevant kind of knowledge is practical, unreflective familiarity. The school does not yet belong to the pupil's environment if he still has to consult a plan and a rule-book. The pupil is not yet 'at home' there. Nothing, of course, prevents a person from reflecting upon, and articulating, practical familiarity; but if he *needs* to do this in order to get about, he is not yet in an environment. Badgers, presumably, are incapable of such reflection and articulation; but what counts is the smoothness and appropriateness of their behaviour in and around the set. For it is in such behaviour that the relevant knowledge must manifest itself. (It may be that animals' lack of a reflective capacity is compensated by a greater intimacy with their environment than humans can achieve. Unable to stand back from, and take critical stock of, their environment, they run no risk of alienation from it.)

A further essential feature of a creature's relation to its environment needs to be brought out. To speak in the language of phenomenology, this relation is an 'intentional' one. An environment, that is, is something *for* a creature, a field of meanings or significance. It is not simply that its environment *matters* to the creature. A badger set, after all, might matter a good deal to a zoology student without thereby belonging to his environment. The point is rather, first, that the items in one's environment are those which are brought into relief, 'lit up', through occupying places within one's everyday practices. For the school pupil, the room is not just a brick-and-mortar structure, but where he goes for history lessons. For the badger, those red berries are not little balls of organic matter, but items to be stored for the winter. The classroom, the berries, have significance within the relevant forms of life; and if they did not, they would not stand out for experience, nor belong in any environment. (Notice that this would not make sense on the 'new' conception, since

The Environment, the natural order, cannot depend for its existence on the practices of certain creatures.)

In calling an environment a field of significance I mean, second, that the items within it signify or point to one another, thereby forming a network of meanings. It is this which confers cohesion, a certain 'wholeness', on an environment, rather as the episodes in a novel belong to a coherent narrative through pointing back and forth. The German philosopher Martin Heidegger describes a person's world – for example, a farm with its equipment, inhabitants and surroundings – as constituting a 'referential totality'. For the various items which belong there – a cow's udder and a milk pail, say – point towards one another and take on significance only as parts of a whole (1962: 90). Animals, too, dwell in fields of significance; the droppings at the entrance to the tunnel indicate a fox, which signifies a threat to the badger's young, whose squealing expresses hunger, which refers the badger to the berries behind that tree, the scent on which means the recent presence of a fox, which indicates . . . etc.

Medieval scholars used to speak of the world as 'the Book of God', an immense collection of 'signs' provided for our benefit by God. On a smaller scale, and without the theological baggage, an environment might be called 'a book of signs' for those whose practical mastery of the right 'language' enables them to 'read it'.

III

It will clarify the account of an environment which I have been articulating if it is compared and contrasted with the rejected accounts in terms of geographical proximity and causal impact. It is important to recognize why these accounts, although inadequate, readily suggest themselves. There are, I think, two reasons why the idea of an environment as a field of significance easily degenerates into these other conceptions.

The first is that there are obvious contingent correlations between a field of significance, what is geographically close and what causally impinges on a creature. Things will tend to matter more to the creature if they affect its sense organs, the spatial range of which is limited. Likewise, if they exert strong causal effects on the creature. Such tendencies, powerful as they are, are only contingent, however. Television and telephone extend the range of the senses, opening up the possibility for distant places to figure in a person's 'intentional' environment. (Here is a grain of truth in the dangerous metaphor of 'the global village'.) And something can be significant for a creature even though, independently of this, it exerts little causal impact upon it – like the nip in the air which heralds for a farmer that winter is approaching.

The second reason why it is easy to slide from the proper account to the others is that there are indeed senses of the terms in which *necessarily* an environment is 'close' to, and replete with 'effects' on, a creature. But this is not geographical closeness, nor causal effects. The closeness in question is more akin to the one a man or woman experiences towards their family and house, even (or especially) when they are on a business trip abroad. And when they are on the trip, their family and home do not cease having effects on them: not because their causal powers extend mysteriously across the ocean, but through remaining in the scheme of the person's concerns, in their memories and future plans. These are, if you wish, figurative senses of 'closeness' and 'effect'. Here, as elsewhere, people are prone to slide into a more literal understanding.

I want now to expand and defend my remark that the current notion of The Environment is, ironically, symptomatic of just that 'scientism' of which 'new' environmental ethicists often complain. One of these writers holds, for example, that 'the biggest philosophical question of our age is whether there might be an alternative to modern techno-scientific thinking', and discovers such an alternative in ecological thought.[4] But ecological accounts of environments, even when their novelty is exaggerated, are very much in the main scientific tradition. They purport, that is, to be 'objective' in the sense of describing the world in terms which are as free as possible from those which register 'subjective' attitudes and feelings. Ideally, the only relations which figure in these accounts are time, space, and causality. Talk of what a thing means to a creature, or of its referring to another thing in a field of significance, has no more place in such accounts than does the terminology of prettiness and ugliness within biological theories. Ecology is as much of a leveller as any other physical science, since the environments of which it speaks are merely instances of general mechanistic processes. Everything that makes an environment special *for* a creature – from the inside, so to speak – is outside this scientific domain. Ecologists, we are told, are more sensitive than other scientists to 'purposes and values of the world'.[5] But such sensitivity is not a credential for ecology and terms like 'purpose' and 'value' no more belong there, in the final analysis, than they do in evolutionary theory.

It might, parenthetically, be worth speculating that the shift in the idea of environment is symptomatic not only of a predilection for a scientific perspective, but of the situation of today's intellectuals. In the eighteenth century, Immanuel Kant managed to pursue a long academic career scarcely ever leaving the confines of his native Königsberg. His successors, today, typically grow up in one place, study in a series of others, work in several institutions, flit from conference to conference and from country to country, and at the press of a button engage with events on the other side of the globe. Theirs is, in David Lodge's sense, a small world; but it is also a very large one, within which fewer and fewer cultivate that unreflective familiarity which, traditionally, marked a person's relationship to an environment. At home everywhere, today's intellectual is at home nowhere in particular. It would be no surprise if his idea of an environment should be The Environment.

IV

Some readers will, by now, be complaining as follows: 'Let's grant that there has been a shift from an earlier notion of environment. Still, the meaning of a word is not inviolate and the fact is that, today, "environment" can refer quite properly to the biosphere and its local parts. This is how it is applied by the "new" ethicists, and nothing you have said discredits their claims.'

It is true that one does not criticize a concept merely by remarking on its novelty. But I want to hold, first, that the tensions, hollowness and vagueness which characterize much 'new' writing are largely due to the distended nature of this new concept, The Environment. Second, the plausibility of the 'new' claims is gained by exploiting the reverberations of the earlier concept. Third, and in consequence, preserving elements of truth in these claims requires a return to that earlier concept.

I now try to substantiate these points – in connection, first of all, with some of the tensions which infect the 'new' thinking. Put schematically, what happens all too often is this: A particular moral concern is lumped alongside others under the heading 'environmental ethics', with two unfortunate results. First, the real tensions which exist between this concern and the others are disguised, the illusion being created that they can all be catered for within a single, 'new' environmental perspective. Second, perhaps because these tensions are dimly perceived, the moral concern in question is granted only a modest place within environmentalist programmes and manifestos – much more modest than the one it would have enjoyed if it had not been absorbed into such an amorphous category.

A very considerable example is provided by certain concerns for animals. Most 'green' manifestos denounce the evils of factory farming and experimentation. These passages are usually found in sections also dealing with other animal issues, like endangered species, under such headings as 'The living environment', which also embrace such topics as conservation of plant life. Now, for an obvious reason, factory farming and experimentation do not belong in the same basket as these other issues. Creatures in the battery or the laboratory have been removed from 'nature', and do not constitute an 'environmental problem' at all in the way that a threatened species of hawk or elm does. That they are animals, indeed living things at all, has become incidental to the functions they serve: the provision of edible matter or of specimens which chemists perform tests on.

For this reason, discussion of the plight of these creatures fits badly with other 'green' issues, which is doubtless why it occupies such a

modest place in most 'green' literature. Factory farms and animal laboratories cannot be catered for by an ethic whose primary vocabulary is one
of 'wilderness', 'nature', 'ecosystem' and the like. Giving prominence to
these two issues would, moreover, soon spoil the veneer of harmony
among environmental concerns. Raising artificially fed calves in a saniti-
zed factory might well be 'environmentally friendly', since they do not
release methane gas into the atmosphere, nor require to graze upon land
which would otherwise be covered by oxygenating forests.

Anyone who is disgusted by current practices of farming and experimentation should therefore regret the subsumption of these under an
awkward umbrella like 'Problems of the environment'. Thus subsumed,
they do not enjoy the salience they deserve.

This is not to say, however, that no environmental thinking is pertinent
to these issues; rather, the relevant sense of 'environmental' needs to be
the one I articulated in sections II and III. Animals, like plants, are of
course in The Environment. That is, they are somewhere, not nowhere.
Crucially, however, animals can also *have* environments. Unlike plants
and lakes, animals ('higher' ones, at least) know their surroundings, and
in just that practical, unreflective, familiar way which makes of them a
'neighbourhood', a field of significance. Things in its environment matter
to the animal through its concerns and purposes; and these things form
a network of meanings in which it knows its way about.

Many evil aspects of our treatment of animals do not require an
environmental perspective, certainly not a 'new' one, in order to be
apparent. But one evil, surely, is to deprive an animal of an environment.
Dr Harry Harlow, a man in the Josef Mengele tradition of research,
used to put young monkeys down long, empty aluminium tubes – 'wells
of despair' – in order to observe the onset of insane depression. This is
extreme deprivation, of course, but in an essential respect the situation
of battery hens and laboratory rabbits is similar. The creature is deprived
of scope for activity, motion even; and it is through intentional activity
that the items around the animal acquire significance and so form an
environment – a precondition, arguably, for its having a life at all. In
some attenuated way, I suppose, the food pellet or syringe matters to
the hen or rabbit squashed up against the wiring of a cage. But these
matter only by causing some automatic response, such as terror, and not
through occupying a place in the animal's scheme of activity, in the life
it leads – for it leads no life, and can have no scheme of activity.

Another important example which, I think, substantiates the points
stated at the beginning of this section is the issue of economic justice
for Third World countries. Here we find the same pattern as in the case
of the animal issues, the masking of tensions under a soothing umbrella
title ('Ethically sustainable development', etc.), and the failure to give
the issue the prominence it deserves. And once again, it could be beneficial to regard the issue from the perspective of the older idea of

environment. Attention might then focus on the cost of 'catching up' with the west in terms of depriving people of environments which are the setting for the traditional activities in which they are 'at home'.[6] But I must leave it to the reader to ascertain if and how Third World issues exemplify the pattern indicated.

V

Next, let me return to the hollow piety of which I accused the 'new' environmentalist vocabulary, in particular to the exhortation to regard nature or The Environment with *reverence*. Taken seriously, this would mean that each of us should regard everything in nature in somewhat the way that, say, a person in Benares views the Ganges. And that implies that the exhortation cannot be taken seriously. If the fashion of expressing reverence towards everything spreads, we shall simply need a new word for what used to be meant by 'reverence'. For it is absurd to suppose that the kind of attitude held by the Hindus to their river could be held by everyone towards everything.

My example of reverence was not random, since that notion is most at home in the context of religion. The object of reverence *par excellence* is a god, but not any old god. It must be one with which people have dealings, and which plays a role in their everyday life: one it makes sense to petition, for example, or to sacrifice to. Thus no one could revere the abstract 'Supreme Being' which the French Revolution substituted for the Christian God. The god must, furthermore, be one on which people feel dependent, not only for their coming into existence, but for the possibility of purpose and flourishing in that existence. The god must be one whose withdrawal of grace or favour is to be feared. ('Reverence' comes from a Latin word for fear.) Allah and Jehovah can be objects of reverence in a way that Bacchus and Brahman cannot: Bacchus because people are not dependent on him in the suitable sense, Brahman because it is not the kind of 'person' which dispenses or withholds favour.

Analogous considerations apply when the object of reverence is in the natural order. A person can only revere what enters into his life, and which belongs, prominently so, within his field of significance. The revered object must, furthermore, induce something akin to the sense of dependence, fear and gratitude which people have towards their gods. Such might be the feeling of an alpine farmer towards the mountain upon whose 'moods' depends not merely his livelihood, but his ability to structure a life for himself and his family, and even his scheme of the value and place of things.

Given these constraints, The Environment cannot be an appropriate object of reverence. The essential point is not that most of The Environment is a very long way from any given person, for an object of reverence need not be geographically close. After the gods, it has perhaps been the sun which has attracted the greatest reverence. This is not, however, the sun considered as a large ball of gas millions of miles away, but the sun as the grower of crops, whose appearance or retreat cheers or depresses the day, and whose various colours and sizes people must learn to read if they are to organize their work and pleasures. Figuratively, the sun is as close to us as anything is. And it is this kind of closeness, an impingement upon everyday concerns and a source for the structuring of one's life, which equips something to be an object of reverence.

There are poets, it will be said, who have expressed a worship of nature as such. But as a reading of one of the greatest, Wordsworth, indicates, such an attitude – when it is not worship of a god deemed to be present in all phenomena – is directed towards particular items in the poet's experience. It was not nature-as-a-whole, but the rocks and 'deep and gloomy wood' by the Wye which 'were then to me an appetite; a feeling and a love' (1971: 23). And the 'natural objects' whose 'gentle agency . . . led me on to feel / For passions that were not my own and think / On . . . human life' were specifically those round Green-Head Ghyll (1971: 68–9).

Sometimes a person is revered, a teacher perhaps – an implication being that theirs is a life to be emulated, a source of lessons for us. Similarly, those who urge reverence towards nature emphasize that it is a model or mistress, a guide for us. The contemporary emphasis is on the study of ecosystems and evolutionary processes as providing lessons for human conduct. Effort is spent, for example, to show that nature is much more an arena of 'co-operation' than of relentless struggle, and to draw an instructive moral from this for human society.

A certain ambiguity in the word 'natural' lends an emotive force to such comparisons between nature and culture; for it refers sometimes to what happens in the unadulterated state of nature, but sometimes to what is 'normal', not 'perverse' or 'artificial'. (One thinks of the carnivore's insistence that, if it were not all right to eat meat, we wouldn't have been equipped with teeth.)

For reasons I have no time to adumbrate, I doubt that ecological and other scientific studies of natural processes can deliver much by way of moral lessons. Human practices have long belonged too indelibly to culture for news about what happens before or beyond culture to provide relevant instruction. There is no harm, perhaps, in speaking of an ecosystem as a 'community' whose 'members co-operate' to preserve the stability of the whole. Provided, that is, we heed John Passmore's reminder that 'this is not the sense of community which generates rights, duties and obligations' (1980: 216).

There is an earlier idea of 'learning from nature' – articulated in Taoism and Zen, by Wordsworth and Thoreau – very different from the current one. According to the earlier idea, the lessons to be learned are obtained not from scientific study, but from everyday acquaintance. Wordsworth, defending time spent in the open air rather than the library, writes:

> One impulse from a vernal wood
> May teach you more of man,
> Of moral evil and of good,
> Than all the sages can.
>   (1971: 33)

A Zen master of archery tells his pupil: 'You can learn from an ordinary bamboo leaf what ought to happen . . . the shot *must* fall from the archer like snow from a bamboo leaf.'[7] The vernal wood and the bamboo leaf spoken of are not botanical specimens, nor is the relevant knowledge of them the kind an ecologist provides. They are the items familiar to anyone who cares to look about him. But how do we learn from such familiarity? Here a second difference from the 'new' approach emerges.

The earlier idea is not that research uncovers deep structural similarities between the natural and cultural, but that nature is an immense source of symbols and metaphors for our lives. Human actions are not in the least similar, scientifically viewed, to drops of water. The point of this favourite Taoist analogy is, rather, to provide an image to hold before us when reflecting upon a certain ideal – that of the fluid, unresisting integration of our actions in the stream of our lives. Nor, when Thoreau calls a lake 'the earth's eye, looking into which the beholder measures the depth of his own nature', is he announcing a remarkable discovery concerning the similarities between water and the human psyche. Rather, meditation on the lake is a spur to the imagination, as when we wonder if the great depth of mountain lakes has its analogy in the depth of character of people in 'mountainous circumstances' (1886: 185, 289).

We are back, in fact, with the idea of nature as a book. As a metaphor in a poem inspires a reader to reflect on one thing through the prism of another, so a natural phenomenon, for the person who 'reads' it poetically, belongs to a vocabulary of symbols which prompt reflections and lend to them a poignancy they would not otherwise enjoy.

VI

My final critical discussion is of the now familiar prescription that we should see ourselves as inextricably part of The Environment, and should develop a sense of oneness or unity with nature. This is the 'new' ethic's antidote to the hubristic anthropocentrism which is blamed for environmental ills. I have already suggested that such prescriptions veer between triviality and falsity according to how they are interpreted. I now want to argue that this is the unsurprising result of operating with a distended notion of environment, and that perception of what is valid in the prescriptions requires, once more, rehabilitation of the earlier notion.

To appreciate the lack of clarity in the theme of unity, it is useful to glance at a recent, popular book by a 'new' writer. In *The End of Nature*, Bill McKibben sounds this theme in familiar style: 'the world displays a lively order. . . . And the most appealing part of this harmony, perhaps, is . . . the sense that we are part of something with roots stretching back nearly for ever' (1990: 67). Such remarks, however, sit uncomfortably next to the central claim suggested by the book's title. This is that man, by intruding to the extent of altering temperature, rainfall, sea levels and the atmosphere, has all but killed nature as traditionally conceived. For nature is what is 'other' than and 'alien' to society and culture. By killing it, moreover, man destroys something he has a deep need for. "We have deprived nature of its independence, and that is fatal to its meaning . . . without it there is nothing but us . . . its loss means sadness at man's footprints everywhere' (1990: 54, 65).

The question surely arises, if it is so important that nature remains 'other' for us, what is meant by the rhetoric of regarding ourselves as but a part of it, at one with it? Nothing could better express *some* kind of integration of man and The Environment, one might suppose, than our capacity to affect nature's most basic processes, so that even rainfall bears a human footprint. McKibben at least owes us an explicit distinction between different senses of 'oneness' which are at work in his account: between the kind which is a 'bad' thing, since it spells the end of nature as 'other', and a 'good' kind which we are urged to cultivate.

It is no wonder that McKibben oscillates between the rhetorics of 'otherness' and 'oneness' when, like many others, he includes under 'environment' everything from the street corner to the stratosphere. Human beings stand in countless relationships to The Environment and its constituents, and it is senseless to ask of these *en bloc* whether they should be relationships of unity within a whole or contrast with what is 'other'. I am writing these words on a ship anchored off the coast of Egypt, impressed by the alien remoteness of the rock and sand which stretch before me. In a few days' time, I shall be equally impressed by the personal, homely feel of the pinewoods where I walk my dog. More theoretically, it is of course right, in certain contexts and by certain criteria, to emphasize man's affinity with other natural and animal life. But sometimes it is equally right to emphasize his uniqueness. Given man's capacities for speech and self-reflection, for instance, slogans to the effect that people are as much part of nature as mountains and fish clearly need to be balanced by ones which recall people's transcendence of the natural.

How does the rhetoric of unity with the environment fare when this is the environment in the sense of an 'intentional' field of significance?

The answer is that it becomes somewhat tautologous, since such an environment is understood as a network with which a creature has a practical, smooth and unreflective familiarity, one it is 'at home' in. The creature is, in this way, part of its environment, though one could as truly say that the environment is part of it. For it is what the creature has, so to speak, made its own, through activities which confer sense on the items in its field, and indeed constitute these items *as* a field.[8]

It follows that it is redundant to exhort a person to feel part of his environment, since this is precisely what he is already 'at one' with. I remarked earlier, however, that it is possible for a creature to lose or be without an environment. So perhaps the demand for unity with an environment might be construed as the appreciation that creatures should be secured environments. This demand would draw its strength from the fear that, in today's world, more and more people (and animals) are bereft of true environments. In part, this would be the fear that it is becoming increasingly difficult to develop that intimate familiarity with a field of signifiance which enables a person to master their world – not in the sense of subduing it, but in a manner akin to mastery of one's native language. Traditional ways are disappearing; people are constantly on the move; historical change is fast and radical. And, in part, it would be the fear that today's world has become too 'disenchanted', as Max Weber put it: too much subject to the canons of scientific understanding for it to be encountered as the repository of meanings and symbols which were once the stuff of daily experience.

VII

The main impetus behind a 'new' environmental ethic has been people's worries about pollution, fossil fuels, radioactive waste and other damaging products of a technological society. In fact, it is far from clear that any 'new' ethic is needed in order to address such issues. It can be argued, certainly, that the real need is for intelligent and robust application of some perfectly familiar moral injunctions. Especially important would be the one discussed at some length by Passmore: duty towards our children and grandchildren, requiring that they be bequeathed a world no less clean and healthy than the one we are willing to tolerate for ourselves. Nor, to take another example, is it obvious that the proper treatment of animals demands a 'new' outlook, as distinct from a clear perception of what is entailed by generally accepted injunctions against causing suffering.

Even if a novel approach to these issues is required, this is not provided, if I am right, by what I have been calling the 'new' ethic. The attitudes it enjoins us to adopt – reverence for, a sense of 'oneness' with, The Environment – are neither feasbible nor coherent.[9] Yet I am sympathetic towards the thought that something is needed, not to replace but to complement the old moral principles of obligations to one's children, avoidance of suffering and the like. These, I suggest, do not have the power to inspire the kind of *mood* favourable to addressing the issues of pollution, global warming, exploitation of animals, etc. The reason that the 'new' ethic does not provide the 'something' in question is that what is required is neither new nor, in any clear sense, a matter of ethics.

Earlier, I emphasized the importance for a creature of having an environment, a milieu which it belongs in and makes its own, which provides it with an arena of significance and in which it can develop the degree of mastery over its life which is appropriate for its species. Far from being 'new', this ideal is so ancient and entrenched that it has rarely been thought necessary to articulate it. That necessity arises only when, as in our technological age, the ideal is endangered. The need, then, is not for the new, but for renewal of something very old. Nor is this ideal exactly an ethical one. (It would be no less appropriate to describe it as pertaining to the *aesthetics* of existence.) A person is not 'morally obliged' to have an environment. This does not, of course, mean that we are dealing with something unimportant. (Rather little of what goes to making up 'the good life' is a matter of rights and duties – affection, for example, isn't.) Indeed, it is hard to know what is more essential to the flourishing of a creature's life than the capacity genuinely to *lead* a life.

What might be the bearing of a self-conscious retrieval of this ancient ideal upon the issues of pollution, etc.? Possibly not very much – though one might add that it is difficult to be any more optimistic about the impact of calls to revere everything in nature. But perhaps the following can be said. The concerns of people conscious of the ancient ideal will begin 'at home', with *their* environments, the networks of meanings with which *they* are daily engaged. And these concerns will be directed at whatever threatens to separate them from their environment, to make their milieu alien. They will be directed, say, at the proposed erection of a factory farm, the squawking and stench from which expel the familiar sounds and smells of their surroundings; or at the planned construction of a motorway which will render impossible the old intimacy between neighbours on opposite sides of the valley.

But these concerns will not remain purely 'local'. While my environmental concerns begin with *my* environment, I recognize that other people (and animals, too) have, or should have, their environments. If I appreciate the importance for my life of a place I know my way about I must appreciate the importance this has for others as well, and I will want to defend their efforts to preserve such places. Perhaps it would be in the emerging, mutually supporting league of little, local pockets of resistance – and not through exhortations to 'global awareness' – that a mood, a psychology, receptive to confronting 'The Problems of The Environment' is nurtured.

NOTES

1 Cooper (1989: 4).
2 Views of the type sketched can be found in any number of places: recently, for example, in several of the essays in Engel and Engel (1990). Needless to say, there are individual writers advocating a 'new' ethic who are not guilty of the monotony, blandness and other faults which I find in so much of the literature – to cite just one example, Sprigge (1984).
3 Thus one queries the inclusion of both 'integration of conservation and development' and 'achievement of equity and social justice' among the 'five criteria' for 'sustainable development' identified by the 1986 Ottawa Conference on Conservation and Development. See Engel and Engel (1990: 8).
4 Ferré (1988: 132).
5 ibid.
6 Documents like the Bruntland Report, *Our Common Future* (World Commission on Environment and Development 1987), and those it has spawned do remark on the importance of local traditions in Third World countries, but without, in my view, taking sufficiently seriously the tensions between this and the aim of 'economic justice'.
7 Quoted in Herrigel (1985: 68).
8 See Sprigge (1984). The idea is a familiar one in phenomenological writings. See, for example, Merleau-Ponty's remark that 'The world is wholly inside me, and I am wholly outside myself' (1962: 407).
9 Perhaps I should stress that my objections are to the kind of 'oneness' urged by contemporary environmentalists. I do not intend my remarks to apply to more metaphysical, perhaps more 'mystical', traditions of 'oneness' – such as that found in the Advaita school of Vedanta thought.

13 Towards a sustainable future

Joy A. Palmer

> Our global future depends upon sustainable development. It depends upon our willingness and ability to dedicate our intelligence, ingenuity and adaptability – and our energy – to our common future. There is a choice we can make.
>
> (WCED 1987)

The World Commission on Environment and Development (WCED) was established in 1983 as an independent body by the United Nations Environment Programme. Its twenty-two members, chaired by Gro Harlem Brundtland, Prime Minister of Norway, aimed to respond to ever increasing concern about the problematic impacts of human activity on the natural resources of the Earth, to examine development problems of our planet and to formulate possible proposals to resolve these problems. The report of the commission, *Our Common Future* (1987), took into account three years of special studies by experts, enquiries, public hearings across the world and consultations with world leaders in business, industry, science, education, politics and development. The report was unanimous in its documentation of both successes and failures in world development programmes. Without doubt, the world approaching its twenty-first century AD sees some positive trends: infant mortality is dropping; human life expectancy in general is rising; global food output is growing faster than world population; and agriculture, medicine and industry share exciting scientific and technological advancements. The good, unfortunately, is outweighed by the bad: forests are disappearing, deserts expanding and soil eroding; the Earth's protective ozone shield is diminishing as air pollution contributes to this global warming; industry and agriculture put ever increasing quantities of toxic substances into our food chains; and large-scale development programmes are failing to narrow the divide between rich and poor.

When the century began, neither human members nor technology had the power radically to alter planetary systems. As the century closes, not only do vastly increased human numbers and their activities have that power, but major, unintended changes are occurring in the atmosphere, in soils, in waters, among plants and animals, and in the relationships among all of these. The rate of change is outstripping the ability of scientific disciplines and our current capabilities to assess and advise.

(WCED 1987)

The scope of the commission's analysis of the global situation is comprehensive and outstanding for its major focus on the concept of sustainable development. Indeed, this concept is now an accepted vital element of world dialogue and debate on environmental issues, largely because of the understanding of its central importance spelled out in *Our Common Future*. As defined by WCED, sustainable development is a dynamic process designed to meet today's needs without compromising the ability of future generations to meet their own needs. It requires societies to meet human needs by increasing productive potential and by ensuring equitable economic, social and political opportunities for all. Sustainable development must not endanger the atmosphere, water, soil and ecosystems that support life on Earth. It is a process of change in which resource use, economic policies, technological development, population growth and institutional structures are in harmony and enhance current and future potential for human progress (WCED 1987).

If a sustainable society is one that satisfies its needs without jeopardizing the quality of life and resources of future generations, then it follows that each generation has a moral responsibility to ensure that future inhabitants of our Earth inherit an undiminished bank of natural resources. It is the task of other chapters in this book to examine this key concept of intergenerational equity and the ethical issues involved. An understanding of the complex concept of sustainability itself is far from straightforward – there are no existing models to consider, and at best one can only build up an academic description, or a vision of what a sustainable society would look like. Key questions immediately come to mind; for example, if the world's rain forests are to be protected from clearance for agriculture, then how is an ever growing population to be fed? If limited reserves of fossil fuels are no longer to be used for our power base, then what will the replacement energy sources be? Key assumptions also have to be taken into consideration; it seems reasonable to assume that continuing reliance on the burning of fossil fuels will cause dramatic and catastrophic changes in the world's climate. New technologies will inevitably be developed. Human population size will continue to increase – a United Nations prediction is that the world will be required to sustain 9 billion people in forty years' time.

The complexity is such that any progress towards sustainable development must take account of a wide range of specific environmental, economic and social goals, and must establish *priorities* within this range of goals. Prioritizing will involve international discussion, debate and cooperation, and the allocation of resources to agreed priorities. A number of recent publications have described global agendas for approaching sustainability and have specified priorities. These include, besides *Our Common Future*, the Worldwatch Institute's *State of the World* (1990b) report, the World Resources Institute's *The Global Possible* (1985) and *Agenda 2000* (1988), a special report by the *Christian Science Monitor*. An analysis of these reports reveals the key categories of environment-related priorities considered essential in any planned movement towards achieving sustainable development. Twelve priorities for concern may be identified, namely:

• Slow population growth.
• Reduce poverty, inequality and Third World debt.
• Make agriculture sustainable.
• Protect forests and habitats.
• Protect freshwater quality.
• Increase energy efficiency.
• Develop renewable sources of energy.
• Limit air pollutants, notably 'greenhouse gases'.
• Reduce waste generation and increase recycling.
• Protect the ozone layer.
• Protect ocean and coastal resources.
• Shift military spending to sustainable development.

In 1989 the governing council of the United Nations Environment Programme (UNEP), representing fifty-eight nations, agreed an agenda of eight priority areas, namely: climate and pollution of the atmosphere; pollution and shortage of freshwater resources; deterioration of oceans and coastal areas; land degradation; biological impoverishment; hazardous wastes and toxic chemicals; management of biotechnology; and protection of health and quality of life (UNEP 1989b).

From an analysis of such reports and supporting data, it is possible to identify key conclusions concerning relationships that exist among important causes of worldwide problems, the impacts of these causes and possible solutions. The five key causes of worldwide environmental problems would appear to be unsustainable population growth, poverty and inequality, unsustainable food production, unsustainable energy use and unsustainable industrial production – wherein the term 'unsustainable' means that these activities endanger or destroy the Earth's natural life-supporting systems. Such causes contribute to a greater or lesser extent to problems such as the extinction of plants and animals, degradation of the land, depletion of non-renewable energy reserves, depletion of freshwater supplies, water and air pollution and the lack of basic human needs such as food, clean water and health care. From an identification of key causes, it is perhaps possible to prioritize related solutions. The slowing of human population growth is arguably the most important single factor in any sustainable development agenda. Data suggest that making agriculture sustainable and using energy sustainably rank as the next most important solutions, whilst others include protecting forests and other habitats, using water sustainably, reducing poverty and reducing the generation of waste.

It is perhaps a sound argument that the drawing of generalized conclusions of this kind is a subjective, arbitrary and almost meaningless exercise in such a highly complex situation involving scientific facts, definitions and values. Nevertheless, if political and economic choices and decisions are to be made on any international agendas, it would seem essential that comparisons are made of alternative policies, possible priorities and respective costs and benefits. For each problem and suggested solution, a wealth of supporting factual data are available. Estimates of the effectiveness of these solutions are inevitable in what can only be a world vision of a sustainable future.

Few would doubt the wisdom of any movement towards sustainability and the immense value in the debate provoked by the Brundtland Commission. The report, however, has not escaped serious criticism, largely focused on the form and definition of development which it encapsulates. Indeed the commission's analysis is based on a certain understanding of development and economic growth wherein we may read a prescription for imposing a western standard of living on all of the world's people, irrespective of their true needs and desires. A number of Third World and environmental organizations and individuals claim that the commission's definition of development is indeed highly contentious. Perhaps the true *causes*, rather than the cures, of the population explosion, ecological devastation, environmental destruction and increasing poverty rest in the western standard of living prescription. Today's dominant development pattern based on western culture and its mechanistic stance takes little account of the diversity of ethical positions, cultures and traditions that actually exist. Furthermore it fails to encapsulate the true complexity and interrelatedness of all of the processes which shape our Earth and maintain its stability.

Vandana Shiva, one of India's leading environmentalists and scientists, and author of Chapter 14 in the present book, has argued that

> The ideology of the dominant pattern of development derives its driving force from a linear theory of progress, from a vision of historical evolution created in eighteenth- and nineteenth-century Western Europe and universalized throughout the world, especially in the post-war development decades. The linearity of history, pre-supposed in this theory of progress, created an ideology of development that equated development with economic growth, economic growth with expansion of the market economy, modernity with consumerism, and non-market economics with backwardness.
>
> (Shiva and Bandyopadhyay 1989)

Shiva's writings aim to redress the balance, articulating the view that diverse cultures of the world ought not to converge with an accepted definition of sustainability modelled on a dominant western development paradigm.

Perhaps the views of Shiva and many other representatives of non-governmental organizations will be seen as one-sided. Certainly it is difficult to accept that environmental damage is the monopoly of western market economies, given the notoriously bad record in this area of the Soviet Union and its erstwhile satellites in eastern Europe.

Nevertheless, Shiva's 'real meaning of sustainability' draws warranted attention to the complexity and interrelatedness of ecological processes and the presumed conflict between management for optimum sustained productivity and the preservation of values inherent in the natural world.

If sustainable development is to be achieved, then the necessary fundamental changes in and modifications to agriculture, energy, forestry and other physical and industrial systems cannot stand alone. Alongside these changes must be a corresponding shift in attitudes and values – in the social, economic, political and moral aspects of human life. Development for a sustainable future must be as much about shifting values as it is about shifting practices. At the baseline of such a shift will be the realization that the Earth in its natural form must be valued – not at the level of how it may be exploited for the support of *Homo sapiens* and our needs and interests, but as an indispensable entity, worthy of value in its own right. A new age of ecological values would replace such negative principles as competition and exploitation with those of respect (for the diversity of life, other cultures and future generations), concern, co-operation and far-sightedness. A new emphasis of principle would be required to underpin the worlds of industry, commerce and leisure. The role of the individual and of personal attitudes, priorities and values cannot be overemphasized in this task of moving towards a sense of collective responsibility for our Earth today and for the future. Such co-operative spirit may highlight the importance of a shared 'ethic of sustainability', and its contribution towards a collective deepening of understanding of the role of human life.

So how might this be achieved? Strategies may involve a number of areas in which progressive practices and ideals may be developed. Key target areas for innovation and changes in policy and practice include national governments, international organizations, business and industry, non-governmental organizations and individual and community initiatives. Relating to these targets are ongoing developments in science and technology, economic policies and natural resource accounting. Vital as all these may be, they cannot be viewed in isolation from the shift in values and ethics already alluded to, and from one other essential concern, namely education.

It should be the entitlement of every individual to have education for, from and about the environment. Its importance was highlighted in the Brundtland Report (WCED 1987), when it called for a 'vast campaign of education, debate and public participation which must start now if sustainable human progress is to be achieved'. This plea reinforces an earlier message of the 1980 *World Conservation Strategy* (IUCN 1980), indicating the essential nature of expanding provision for environmental education in attempts to change 'the behaviour of entire societies towards the biosphere'.

On the surface at least, a review of trends in environmental education provision (in the United Kingdom) over the past few years suggests grounds for optimism. The subject has been well established on the curriculum map of schools for some three decades, gaining increased significance since the first intergovernmental conference on environmental education held in Tbilisi, USSR, in 1977. This major event and publications based on it continue to provide the framework for the development of environmental education in the world today. Probably the most significant event of the more recent past is the 1988 agreement of the Council of Education Ministers of the European Community 'on the need to take concrete steps for the provision of Environmental Education . . . throughout the Community'. The council then adopted a resolution on environmental education to that end.

The publication of the Department of Education and Science's Curriculum Matters 13 booklet, *Environmental Education from 5–16* (DES 1989), was well timed to follow this resolution. Of even greater significance has been the adoption of environmental education as a cross-curricular theme in the national curriculum for schools (National Curriculum Council 1990) and its content is reflected in the programmes of study and attainment targets of a number of the curriculum's core and foundation subjects.

Notwithstanding such encouraging signs, there still remains a wide chasm between the extent and quality of environmental education provision and the need for the integration of successful programmes with coherent policy. The Council for Environmental Education (1990) articulates a case for a necessary alliance between environmental education and environmental policy, reinforcing that, whilst potential is there, 'it is not realized because environmental education is still widely undervalued, under-resourced and inadequately understood'.

In an ideal situation, there would be immediate and massive expansion of environmental education at all levels, bearing in mind that education is no short-term process and that time is of the essence to the future of our planet. Such education must take account not only of the practical dimensions of sustainable lifestyles and development issues, but equally of differing values and ethical implications. A holistic understanding of natural systems and the place of human endeavour and concern within these would seem essential to the success of any policies aimed at leading the world in the direction of a sustainable future.

14 Recovering the real meaning of sustainability

Vandana Shiva

How quickly words lose their meaning in our times. Take the case of 'sustainability'. It is derived from 'sustain', which means support, bear the weight of, hold up, enable to last out, give strength to, endure without giving way. Sustainability is a term that became significant in development discourse in the 1980s because four decades of the development experience had established that 'development' and its synonym 'economic growth', which were used to refer to a sustained increase in per capita income, were unsustainable processes. Development was unsustainable because it undermined ecological stability, and it destroyed people's livelihoods. 'Growth with equity' and 'growth with sustainability' were attempts to legitimize and perpetuate economic growth in a period of doubt. Economic growth had promised to create abundance. It had promised to remove poverty. Instead, by causing the destruction of livelihoods and life-support systems in the Third World, growth itself became a source of poverty and scarcity. While the 1970s focused on the growing polarization and inequality that went hand in hand with economic growth, in the 1980s the focus shifted to the issue of 'sustainability'. For ecology movements in the Third World, the two issues are usually non-separable – justice and sustainability, equity and ecology are inherently linked in a situation in which the majority of people are excluded by the market economy and continue to draw sustenance from nature's economy. Development had been based on the growth of the market economy. The invisible costs of this development have been the destruction of two other economies, of nature's processes and people's survival. The ignorance or neglect of these two vital economies of nature's processes and people's survival has been the reason why development has posed a threat of ecological destruction and a threat to human survival which have, however, remained 'hidden negative externalities' of the development process.

While trade and exchange of goods and services have always been present in human societies, these were subjected to nature's and people's economies. The elevation of the domain of the market and man-made capital to the position of the highest organizing principle for societies has led to the neglect and destruction of the other two organizing principles of ecology and survival which maintain and sustain life in nature and society.

Modern economics and concepts of development cover a negligible portion of the history of human interaction with nature. Principles of sustenance have given human societies the material basis of survival over centuries by deriving livelihoods directly from nature through self-provisioning mechanisms. Limits in nature have been respected, and have guided the limits of human consumption. In most Third World countries large numbers of people continue to derive their sustenance in the survival economy which remains invisible to market-oriented development. And all people in all societies depend on nature's economy for survival. The market economy is not the primary one in terms of the maintenance of life. When sustenance is the organizing principle for society's relationship with nature, nature exists as a commons. It becomes a resource when profits and capital accumulation become the organizing principles and create an imperative for the exploitation of resources for the market. Without clean water, fertile soils and crop and plant genetic diversity, human survival is not possible. These common resources have been destroyed by economic development. This has created a new contradiction between the economy of natural processes and the survival economy, since those pushed out by development are forced to survive on an increasingly eroded nature.

The organizing principle of economic development based on capital accumulation and economic growth renders valueless all properties and processes of nature and society that are not priced in the market and are not inputs to commodity production. This premiss very frequently generates economic development programmes that divert or destroy nature's and people's base for survival. While the diversion of resources, like diversion of land from multi-purpose community forests to monoculture plantations of industrial tree species, or destruction of common resources, or diversion of water from production of staple food crops and drinking water needs to cash crops, is frequently proposed as a programme for economic development in the context of the market economy, this creates underdevelopment and scarcity in the economies of nature and survival. Having caused the erosion of nature's nature and people's nature, the market is now being proposed as a mechanism for ecological renewal.

While development as economic growth and commercialization are now being recognized as being at the root of the ecological crisis in the Third World, they are paradoxically being offered as a cure for the ecological crisis in the form of 'sustainable development'. The result is the loss of the very meaning of sustainability. The ideology of sustainable development is, however, limited within the limits of the market economy. It views the natural resource conflicts and ecological destruction as separate from the economic crisis, and proposes solution to the ecological

The real meaning of sustainability   189

crisis in the expansion of the market system. As a result, instead of programmes of gradual ecological regeneration of nature's economy and the survival economy, the immediate and enhanced exploitation of natural resources with higher capital investment gets prescribed as a solution to the crisis of survival. Clausen, as the president of the World Bank, recommended that 'a better environment, more often than not, depends on continued growth' (Goldsmith 1985: 2). In a more recent publication Chandler (1986) further renews the argument in favour of a market-oriented solution for the ecological problems and believes that concern for conservation can only come through the market.

Speth (1989) believes that economic growth is imperative and 'only technology can save us'. There is, however, another view; at the World Wilderness Congress in 1987, Oren Lyons said: 'I heard today that economic growth is a necessity and conservation is a consideration of importance . . . we disagree. Conservation is life and economic growth is a matter of interpretation.'

I would like to understand why questioning the sanctity of growth is still considered taboo among development agencies and development experts. I will also try to follow Oren Lyons in showing that preserving the sanctity of development and growth through 'sustainable development' is based on a false interpretation of sustainability. Economic growth takes place through the over-exploiting of natural resources which creates a scarcity of natural resources in nature's economy and the people's survival economy. Further economic growth cannot help in the regeneration of the very spheres which must be destroyed for economic growth to take place. Nature shrinks as capital grows. The growth of the market cannot solve the very crisis it creates. Further, while natural resources can be turned into cash, cash cannot be turned into nature's ecological processes. Those who offer market solutions to the ecological crisis limit themselves to the market, and look for substitutes to the commercial function of natural resources as commodities and raw material. However, in nature's economy the currency is not money, it is life.

Increased availability of financial resources cannot regenerate the life lost in nature through ecological destruction. An African peasant captured this in his statement: 'You cannot turn a calf into a cow by plastering it with mud.'

The task of ecological regeneration and recovery of sustainability is, metaphorically, the task of allowing the calf to grow into a cow. The pseudo-sustainability of 'sustainable development' is the equivalent of plastering it with mud. The false notion of sustainability is based on three flaws. The first is assigning primacy to capital. The second is the separation of production from conservation, making the latter dependent on capital. The third error is assuming substitutability of nature and capital.

Development has been based on assigning supremacy to the market economy, and to its organizing principle based on profits and capital accumulation. 'Sustainable development' preserves the false assumption that the economy as defined by capital and markets is primary and more basic to human well-being than nature's economy of self-renewal or people's economy of sustenance. The latter are considered 'primitive', 'backward', 'stagnant', secondary, and may therefore be destroyed for the sake of development.

In the market economy, the organizing principle for relating to nature is the maximization of profits and capital accumulation. Nature and human needs are managed through market mechanisms. Demands for natural wealth are restricted to those demands registering on the market; the ideology of development is in large part based on a vision of bringing all of nature's products into the market economy as raw material for commodity production. When these resources are already being used by nature to maintain its renewability and by people for providing sustenance and livelihood, their diversion to the market economy generates a condition of scarcity for ecological stability and creates new forms of poverty for people.

Traditional economies based on principles of providing sustenance with a stable ecology have shared with industrially advanced affluent economies the ability to utilize nature to satisfy basic vital needs of food, clothing and shelter. The former differ from the latter in two essential ways. First, the same needs are satisfied in industrial societies through longer technological chains requiring higher energy and resource inputs and higher creation of waste and pollution, while excluding large numbers of people without purchasing power and access to means of sustenance. Second, affluence and overproduction generate new and artificial needs and create the impulse for over-consumption, which requires the increased exploitation of natural resources. Traditional economies are not 'advanced' in the sphere of wasteful consumption, but as far as the satisfaction of basic and vital needs is concerned, they are often what Marshall Sahlins has called 'the original affluent society'. The needs of the Amazonian tribes are more than satisfied by the rich rain forest; their poverty begins with its destruction. The story is the same for the Gonds of Bastar in India or the Penans of Sarawak in Malaysia.

The paradox and crisis of development arise from the mistaken identification of culturally perceived poverty with real material poverty, and the mistaken identification of the growth of commodity production as providing better human sustenance. In actual fact, there is less water, less fertile soil, less genetic wealth as a result of the development process. Since this natural wealth is the basis of nature's economy and the people's survival economy, their scarcity is impoverishing people in an unprecedented manner. The new impoverishment lies in the fact that nature which supported their survival is being exploited by the market economy while they are themselves excluded and displaced by the expanding control of man-made capital over nature's and people's life through the process of development.

The real meaning of sustainability would make it clear that nature's economy is primary, and the money economy is parasitic on it. It would also make it evident that the growth of markets and production processes at the cost of nature's stability is at the root of the crisis of sustainability. Sustainability therefore demands that markets and production processes be reshaped on the logic of nature's returns, not on the logic of profits, capital accumulation and returns on investment. Conservation has thus to be a basis of production and exchange. 'Sustainable development', however, protects the primacy of capital. It is still assumed that capital is the basis of all activity. The preservation of the primacy of capital creates a dualism between 'conservation' and 'development'. Traditionally, in the Third World, the three economies of nature, people and markets have always coexisted. Natural forests have been wild regions; they have also been homes of the forest dwellers, and they have supplied markets with forest produce. 'Development' creates mutually exclusive categories of 'human settlements', 'wildlands' and 'productive forests'. 'Sustainable development' continues that trend. People are pushed out of forests to have wildlife and biosphere reserves. While diversity is protected in these set-asides, 'productive forests' are based on the traditional logic of uniformity and market orientation, symbolized by monocultures of fast growing species. It matters little that, in nature's economy and people's economy, commercial fast growing species are often unproductive or even counterproductive.

Production is untouched by ecological principles. Conservation is added on as a mutually exclusive activity. The assumption that 'continued economic growth is essential for ecological recovery' arises from this artificial separation of development from conservation. Conservation is reduced to 'wilderness' management, which needs financial inputs. On this view, people's self-provisioning economies do not fit into 'conservation' activity because they do not fit into the category 'wild', since nature is defined as free of humans. Since this commercial approach to conservation is conceptualized as dependent on finances, and increased financial resources can only be generated through economic growth, it is assumed that economic growth is an imperative for conservation. The split between conservation and development, the dependence of conservation on economic growth, are elements of a false concept of sustainability. True sustainability demands that ecological principles be incorporated into production processes to reshape them. Conservation has to be a *basis* and foundation of production. It cannot be an addendum.

The third fallacy which preserves the sanctity of economic growth in a period of development-induced ecological disasters is the assumption of the non-destructibility of capital and the substitutability of capital and nature. Development was based on the assumption that man-made capital is a substitute for nature's capital, and that flows of cash and currency can replace nature's flows and processes. Georgescu-Roegen (1974: 21) has observed how the substitutability assumption of modern economies has blocked the perception of ecological destruction:

> The no deposit no return analogy benefits the businessman's view of economic life. For, if one looks only at money, all one can see is that money just passes from one hand to another; except by regrettable accident it never gets out of the economic process. Perhaps the absence of any difficulty in securing raw materials by those countries where modern economics grew and flourished was yet another reason for economists to remain blind to this crucial economic factor. Not even the wars the same nations fought for the control of the world's natural resources awoke the economists from their slumber.

The economists' slumber continues in spite of the loud alarms of the ecological crisis. Solow, the 1987 Nobel Prize winner in economics, holds that production and growth can completely do away with exhaustible natural resources and resources exhaustion is not a problem. It is alleged that 'the ancient concern about the depletion of natural resources no longer rests on any firm theoretical basis'. This belief of modern economics is based on its unquestionable faith in modern western science. As Solow (1974: 11) states: 'If it is very easy to substitute other factors for natural resources, then there is, in principle, no problem. The world can, in effect, get along without natural resources, so exhaustion is just an event, not a catastrophe.'

Agencies like the World Bank and the World Resources Institute look towards financial and technological fixes to the ecological crisis with the same optimism. However, further economic technological growth, which acts against nature's logic, must generate accelerated unsustainability. It cannot become a source of sustainability. Life can be destroyed to create capital. This destruction will be more efficient and more rapid with the new biotechnologies which are often offered as miracle cures to the ecological crisis. But accumulated capital cannot recreate the life once destroyed.

There are, quite clearly, two different meanings of 'sustainability'. The real meaning refers to nature's and people's sustainability. It involves a recovery of the recognition that nature supports our lives and livelihoods and is the primary source of sustenance. Sustaining nature implies maintaining the integrity of nature's processes, cycles and rhythms. There is a second kind of 'sustainability', which refers to the market. It involves maintaining supplies of raw materials for industrial production. This is the conventional definition of 'conservation' as making available sustained yields of raw material for development. And since industrial raw materials and market commodities have substitutes, sustainability is translated into substitutability of materials, which is further translated into convertibility into profits and cash.

Sustainability in nature involves the regeneration of nature's processes and a subservience to nature's laws of return. Sustainability in the market-place involves ensuring the supplies of raw material, the flow of commodities, the accumulation of capital and returns on investment. It cannot provide the sustenance that we are losing by impairing nature's capacities to support life. The real meaning of sustainability needs to be based on the insights of the native American elder who indicated that money is not convertible to life: 'Only when you have felled the last tree, caught the last fish and polluted the last river, will you realize that you can't eat money.'

15 Technological risk: A budget of distinctions

Mark Sagoff

In 1983, the US National Academy of Sciences issued a report urging public officials to distinguish between the assessment and the management of risk or, as the academy put it, between the scientific and the political aspects of regulatory policy. The academy report distinguished, in other words, between judgements of fact and judgements of value. Judgements of fact attempt to describe the way the world is. Judgements of value distinguish right from wrong, good from evil, and make recommendations about what we ought to do.

The report advised scientists to avoid introducing considerations of value, in so far as possible, into the process of measuring the magnitude of a risk, that is, the process of measuring the severity of a potential harm multiplied by the probability that it will occur. When scientists maintain this 'objective' spirit in risk assessment, so the report concluded, policy makers may regulate technologies efficiently, which is to say, in ways that maximize the benefits and minimize the costs that flow from the risks associated with those technologies.[1]

FACT/VALUE

The authors of the academy report were not naive; they understood that the distinction between risk assessment and risk management, while useful, is hardly a hermetic one. Officials and others who rely on this distinction generally recognize that 'policy' or 'management' decisions enter even the most rigorous and objective attempt to quantify risk.

The 'management' questions that affect quantified risk assessment include, for example, which hazards to measure, which tests to use, which tumors to count, which extrapolations to countenance, what sort of uncertainty to tolerate, and so on, in estimating the probability of harm. As William Ruckelshaus has said: 'Risk assessment is necessarily dependent on choices made among a host of assumptions, and these choices will inevitably be affected by the values of the choosers, whether they be scientists, civil servants, or politicians' (1985: 28).

No one, so far as I know, argues that risk assessment can or should be an entirely 'value-free' discipline (whatever that could mean) or that there is just one legitimate method to determine the facts. The consensus, rather, seems to be that risk assessment may remain perfectly scientific as long as it brings its normative choices and assumptions to the surface and deals with them in an open and critical spirit. 'Although we cannot remove values from risk assessment,' Ruckelshaus summarizes, 'we can and should keep those values from shifting arbitrarily with the political winds' (1985: 28).

Scientists involved in the assessment of risk tend to be candid about the value judgements they must make in order to gather and interpret data. As two commentators write:

> Whatever the suspected hazard in question, estimating its dangers involves essentially four lines of investigation: defining the conditions of exposure, identifying adverse effects, determining the probable relationships between exposure and effect (such as dose/response), and calculating overall risk. Each of these tasks obviously calls for information; each also calls for judgement – in fixing the scope of inquiries, conducting investigations, interpreting findings, determining their weight.
>
> (Gilette and Krier 1990: 1062)

Opponents and proponents of new technologies – for example, nuclear power and genetic engineering – might come to their positions for aesthetic, moral, cultural, religious, economic or political reasons that may have little to do with size of the risk these technologies pose. Yet, rather than articulate these legitimate concerns, proponents and opponents may find it easier to argue over the accuracy of a risk assessment. Because risk assessments require interpretive and evaluative judgements, they are easy to question. Accordingly, the assessment of risk can become a stalking horse for other equally important considerations.

To see what considerations other than the size or severity of a risk may attract people to a technology or repel them, I should like to assume that scientists know how to assess risks in an objective way. I mean by this not that they can keep from making value choices altogether, but that they may acknowledge and be consistent in the normative choices they must make – for example, when they adopt one protocol rather than another to extrapolate animal data to human populations. Let us assume, then, that policy makers get the best information they can get, however uncertain it may be, about the magnitude of risk a novel technology may pose. The question I wish to discuss then arises. What should risk managers, or regulatory agencies, do with that information?

Risk management, according to Ruckelshaus, ideally 'is based on such factors as the goals of public health and environmental protection, relevant legislation, legal precedent, and the application of social, economic, and political values' (1985: 28). I wish to discuss these factors. I want to examine some of the reasons we may judge one but not the other of two equally severe hazards to be acceptable, within the social, cultural, political or moral context in which they arise.

When we investigate technological risk within a political and cultural context, we usually think in terms of a group of distinctions which, presumably, capture the values on which risk management decisions ought to be based. This budget of distinctions typically includes the following pairs: voluntary, involuntary; autonomous, heteronomous; public, personal; natural, artificial; fair, unfair; and many others. The purpose of this chapter is to take a philosophical glance at some of these distinctions in order to see if they can help us determine the right regulatory response to those technological risks the magnitudes of which have been reasonably well identified. I want to understand something about the *acceptability* of technological risks, in other words, quite apart from disagreements about the actual magnitude of those risks.

VOLUNTARY/INVOLUNTARY

In 1969, Chauncey Starr pointed out: 'We are loath to let others do unto us what we happily do unto ourselves' (1972: 30). Starr divides social activities into two general categories: 'those in which the individual participates on a "voluntary" basis and those in which the participation is "involuntary," imposed by the society in which the individual lives' (1969: 1233).[2]

What constitutes the difference between voluntary and involuntary risks? Can we make enough sense out of this distinction to apply it in regulating technological hazards?

We may answer this question by distinguishing two kinds of freedom or voluntariness. First, we may speak of freedom in a causal or metaphysical context. An act is voluntary in this sense if the agent knows what he or she is doing and could have done otherwise. Thus, as long as an agent could have chosen not to take a certain risk, and yet knowingly takes it, his or her choice is free in a metaphysical sense.

Second, we may speak of freedom in a moral rather than in a metaphysical sense. A person is free morally only when he or she acts on values or principles which, all things considered, represent his or her reflective conception of the good life and the values that enter that life. A person acts voluntarily in this sense, moreover, only if the values and principles which motivate him or her are autonomous, which is to say, they arise from his or her conception of himself or herself as a free and equal agent.

Starr and other policy analysts conceive of voluntariness in the first or metaphysical sense. Accordingly, they assimilate voluntary risks to risks we knowingly take in circumstances in which we could have done otherwise. On this analysis, a gamble is voluntary if the agent knows the risks he or she takes and is not forced to take them. As long as a person has a choice and knows what it is – as long as he or she is informed and could have declined the bargain – that person chooses voluntarily, on this account, and even consents *ex ante* to the consequences, however brutal they may be.

Kip Viscusi, for example, argues that 'market-traded risks are the result of individual choices' (1983: 1). Viscusi believes that workers should be informed of the risks they take, e.g. in exchange for a higher wage. An informed worker who takes a risky job for a higher wage, better job security or whatever, according to Viscusi, voluntarily accepts those risks. The worker increases his or her 'expected utility' by taking the risk, moreover, whatever may befall him or her as a result.

Thus, Viscusi concludes that humanitarian regulations making the riskiest work less risky, however many lives these policies save, are misguided because they decrease the expected utility of workers who are willing to take those jobs. Statutes of this sort – for example, child labour laws – limit the range of voluntary choices. 'While perhaps well-intended, intervention based solely on these [i.e. humanitarian grounds] will necessarily reduce the welfare of the poorer workers in society, as perceived by them' (Viscusi 1983: 80).

This conception of the 'voluntary' seems too broad to be interesting. People who are mugged, murdered, hit by drunk drivers, etc., often have 'gambled and lost', in the sense that they choose to live in the city, to walk across a street or to go out at night, even though they are apprised of the risks they take. Just as a worker increases his or her expected utility by going down into an unsafe mine, so you, too, increase your expected utility by jogging in the public park. Accordingly, the reason Viscusi might deplore the 'intervention' of mine inspectors – it limits the kinds of risks people can take – seems to apply equally to the 'intervention' of park police.

No one forces you to jog down a street (you might get mugged or struck down by a speeding vehicle); similarly no one forces a miner to work in an unsafe mine. Is it wrong, then, for the government to intervene on the side of safety, say, by regulating the speed at which cars are driven? If not, why is it wrong for the government to 'intervene' (usually after intense lobbying by the miners' unions) to regulate the safety of mines?

Shall we say that the government ought not to regulate risks of any kind, by arresting muggers, say, since the victims were informed of those risks and still took them? Shall we say that these victims, because they walk in the park or drive at night, knowing the risks, consent to being mugged, robbed or killed? Shall we comfort the victim by telling him that because he chose to cross the street knowing that a drunk driver could come along, his expected utility is increased even though he is maimed or dead?

The metaphysical voluntariness of risk, as this *reductio* suggests, seems irrelevant to regulatory policy. The government must establish rules and conditions under which threats and offers of certain kinds are and are not tolerated. It is only within these rules and under these conditions – on which any moral conception of freedom or voluntariness relies – that we can speak of freedom to choose in a sense that is significant for social regulation.

When we divide 'voluntary' from 'involuntary' risks, then, we find that there are two ways we can understand this distinction. We may say, first, that anyone takes a risk 'voluntarily' who understands the relevant probabilities and could have refused the bargain. This conception of the voluntary, however, amounts to little more than the assertion that any unforced choice we make, as long as we are reasonably informed about it, is voluntary or free, however brutal the circumstances, however ugly the results. This idea of liberty defines the Hobbesian state of nature. As a regulatory principle, it is a recipe to ensure that human life will be mean, nasty, brutish, solitary and short.[3]

Second, we may mean that a person's choice is voluntary if it arises in circumstances in which that agent acts on values and in circumstances that express rather than mock his or her ideals and aspirations. We should then analyse concepts like freedom and voluntariness in terms of more basic moral notions – for example, concepts of autonomy and equality. We should then see that freedom in a morally important sense attaches not to any choice an informed person may make but only to those choices the circumstances or conditions of which are themselves chosen under a general conception of justice and social co-operation. These values, which we may associate with self-determination in a morally interesting sense, have nothing to do with the freedom Viscusi defends, namely, the freedom of the poor and rich alike to sleep under the bridges of Paris.

Accordingly, one is drawn to the conclusion that the 'voluntary/involuntary' distinction taken by itself cannot help us divide acceptable from unacceptable risks – for example, in the environment and in the workplace. We must look further into the circumstances in which choices arise to find out which risks are 'voluntary' or 'consensual' in a normatively interesting sense. We look to other moral distinctions, therefore, in terms of which we can say that risks, however small or great, are acceptable or unacceptable from a social, moral or political point of view.

CONSUMER/CITIZEN

People make decisions about risky technologies in two basic ways: first, as consumers entering and exiting from markets and, second, as citizens participating in the political process. And it is very often the case – indeed, it seems more often than not the case – that people will behave one way as consumers with respect to risk and in quite a different way as citizens.

Consumers who enter and exit from markets act essentially in the same way as pedestrians who cross a street. They act as individuals pursuing whatever personal goals they may have within the options and conditions available to them. Citizens participating in a political process, in contrast, act as members of a community, making decisions for the community as a whole, rather than only for themselves as individuals. They do not resemble pedestrians who act just for themselves. Rather, they resemble the transportation planning board that lays out and builds the roads everyone will use.

Such a planning board must agree on a plan; this requires deliberation, reason, argument, political give-and-take and so on. These are the virtues of discussion and compromise on which freedom in a democratic society is based. In a political context, social values and options are not 'given', as they might be in a market, but are to be chosen through the representative processes of democratic government. Thus, we can contrast choices individuals make in markets, on whatever motives they may have, with choices they make after open debate and deliberation in majoritarian democratic institutions.

The question now arises in which context a person should decide the risks he or she takes. Should the person decide in a market or in a political framework? Should regulatory policy respond to the preferences one seeks to satisfy as a consumer in markets or the views and beliefs one vindicates as a citizen through the political process?

This difference between 'consumer' and 'citizen' behaviour may puzzle those in regulatory agencies who need to determine whether the public finds a certain risk acceptable or not. People who move to and live in Los Angeles, for example, may seem fairly indifferent to air quality, since they pay high housing costs to reside in by far the most polluted region of the United States. Yet the citizens of Los Angeles and California generally support with political fervour the most stringent air quality regulations. Where, then, do they stand? Should we base risk management on their market or on their political behaviour?

A person who frequently neglects to use her seat belt may support laws that require drivers to buckle up for safety. Someone who lobbies for the strictest pesticide safety standards may forget to wash the peaches she eats. My parents, now in their eighties, demonstrate against nuclear power, though as consumers they depend on that beleaguered industry. People who are hooked on cigarettes may advocate governmental efforts to discourage smoking to keep others from sharing their awful addiction. Judging from consumer behaviour, we should infer that those people favour 'smokers' rights'; their political behaviour, however, would lead one to the opposite conclusion.

In crusades to end child labour, to make the mines and railroads safe, to control the adulteration of food and so on, individuals who may have sent their children to work in unsafe mines so that they could purchase adulterated food found a way to express values quite different from those social scientists could infer from their market behaviour. Through the political process, citizens are able to change the conditions under which they then make consumer choices. Where should we say their true values lie? Should we measure the value of safety, in other words, in relation to the choices they make as individuals in markets or the choices they make as citizens through the political system?

Many economists have an answer to this question. They believe that people express their true values through their market choices. According to the view, government 'intervention' thwarts or skews those choices, leaving people less well off than they might otherwise have been. Thus, Martin Bailey has argued that 'The most direct evidence of the amount people are willing to pay for their own safety comes from the job market, which offers a variety of working conditions with various degrees of personal risk.' Workers generally demand more money to do riskier work. 'For such workers, the wage differential precisely measures their willingness to pay for increased safety' (Bailey 1980: 33, 31).

Bailey reasons that the value a person places on risk equals the additional increment of income that would leave him or her indifferent between two jobs that differ only in so far as one is more risky than another. 'To convert this information into an amount per life lost or per death avoided,' he writes, 'divide the extra wage by the extra risk' (Bailey 1980: 33).

Apply the formula that Bailey suggests to *unregulated* markets, for example, in turn-of-the-century America or in some Third World countries today, and the revealed value of a life saved or death avoided plus 75 cents might buy a cup of coffee. Legal intervention – statutes backed by the political force of organized labour together with a sympathetic citizenry – moved markets to the point at which workers could make bargains that reflect a 'value' for life that is not unconscionable.

In the United States a century ago the most risky jobs – railroad workers, for example, faced a 10 per cent incidence of death or incapacitating injury – were among the least well paid. Since wages went down as the probabilities of injury and fatality went up, one might infer a negative value for safety. The safety-conscious worker had little to bargain for. He or she was free to choose between getting black lung from mining or byssinosis by working in a textile plant.

Bailey arrives at a value for life having a low bound between $170,000 and $300,000 and a high bound between $584,000 and $715,000. These amounts, which Bailey infers from 1980 data, do not affront our moral intuitions. They reflect, however, labour markets that have been stringently regulated through more than a century of political regulation. It is absurd, then, to read these figures as if they express a true *laissez-faire* market estimate of worker willingness to pay for safety and health. To get *that* sort of number, which is likely to be very low, you must go to where there is no government intervention – to the dark satanic mills of eighteenth-century England.

We are now in a position to answer the question whether public officials should estimate the value of safety in terms of the behaviour of the citizen or of the consumer. The answer I would give is this. We cannot base regulation on the performance of consumer markets. Market data are themselves thoroughly informed by, indeed determined by, the political regulation of risk. These data tell us what regulation has accomplished so far – not what it has left to accomplish.

At the same time, laws like the Occupational Safety and Health Act and the Clean Air Act may go too far in the opposite direction. They call for zero risk – for conditions that ensure that no one is ever hurt – and they sometimes impose an adequate margin of safety on top of that. These statutes may express the ideals and aspirations of a political community, but to implement them rigorously we might have to bring the economy to a screeching halt.

The appropriate compromise must lie somewhere between the reductionism of markets and idealism of politics, between what we seem willing to pay for as consumers and what we nevertheless insist upon as citizens. Earlier, we asked whether the individual should consent to risks through consumer or citizen choice. The answer may be both; the finer-grained decisions may be appropriate to markets, but some of the basic conditions of choice may be appropriately decided in the political process.

The institutions of democracy, at least in principle, allow people to regulate as citizens the markets in which they then compete and co-operate as individuals. As we shall now see, this access to political power increases not just the freedom with which people make choices but also the autonomy they enjoy in making those choices. I shall argue in the next section that autonomy may be as important a concept as voluntariness in determining the acceptability of risk.

AUTONOMY/HETERONOMY

Autonomy has less to do with the outcome of a decision one makes than with the circumstances under which one makes it. A decision may be described as autonomous in so far as the conditions in which it arises come under the control of the agent or at least are not controlled by anyone else. We may think that freedom has to do with a person's welfare in so far as it involves getting what one wants or doing as one likes. 'Autonomy, by contrast, does *not* refer to the degree of satisfaction, but to the *character of the circumstances on which the satisfaction is dependent*' (Kennedy 1973: 371).

Autonomy depends, first, on the control one exercises over one's desires and over the options among which one makes one's choices. Autonomy increases, moreover, in so far as a person is not dependent upon others for the means necessary to achieve his or her ends. Autonomy does not depend, then, simply on a person's acting on wants, desires or interests he or she happens to have, but on the nature of those interests, their origin in the self and their order and structure with respect to general goals and principles which a person affirms and is willing to defend.

Goals and principles are autonomous, in other words, when they arise from a consistent conception of the good life or from a person's idea of himself or herself as a free, equal and rational person. Thus, the mere fact that an action satisfies given desires or is cost-beneficial from an individual's point of view does not show that it flows from the autonomy of that individual. We should have to know how the action and the desires it satisfies relate to the individual's conception of himself or herself as a free and equal person within a larger social order.

Many recent writers on social regulation criticize the failure of agencies to compare risks associated with new technologies (e.g. nuclear plants) with hazards associated with old technologies (firewood, coal) they replace. These writers note that many new technological risks do not add to but substitute for old and familiar hazards. According to this argument, 'screening' new technologies for safety – in so far as it prevents these technologies from developing – is often counterproductive, because it tends to increase rather than decrease overall levels of risk (Huber 1983).

The regulation of technological risk, according to this argument, may transfer dangers from one segment of society to another, rather than provide a net improvement in safety. Even if this is true, however, we should ask whether these transfers are justified, because they result, for example, in an overall increase in personal autonomy.

One may concede for the sake of argument that nuclear power is safer and pollutes less than the use of wood stoves to heat houses. Yet one can reply that many individuals find the family hearth, but not a vast and alien nuclear technology, conducive to their autonomy, self-sufficiency and independence. This is because people are better able to control these smaller technologies; they may gather their own wood, while the larger technologies seem to transfer control from individuals and from democratic institutions to a distant and not-quite-accountable professional elite.

Large-scale complex technologies may increase welfare in so far as they provide consumers more goods at lower prices. Yet they may decrease autonomy in so far as they make individuals dependent on larger and larger systems of production they cannot control or understand. The individual may therefore experience technological risk not as a cost of production but as an additional intrusion on or infringement of autonomy. Accordingly, he or she will resent the risk and deplore analyses which would 'price' it simply in terms of the direct harms it may threaten.

Novel technologies may not satisfy our wants and preferences so much as change them – we adjust our lives to the available technologies as much as the other way round. The risks associated with these new technologies, therefore, are not, as a rule, essential to activities in which we express our conception of the good life or our idea of ourselves as free, equal and rational persons. Technological risks, indeed, are likely to deprive us of autonomy, in so far as they remove from our control the conditions under which we satisfy our interests and determine our values, just as the technologies themselves seem to undermine that autonomy.

The mountain climber, the soldier, the astronaut and others who choose to look death in the face do not accept any old risk but just those risks they find harmonious with their autonomy. Likewise, in less dramatic ways, we may find rather large, familiar risks acceptable if we sense that we control the circumstances in which they arise. But we may object even to small risks if their presence and outcome lie entirely outside our own control.

For this reason, it is perfectly rational for society to regulate risks associated with new, large-scale technologies more strictly than risks associated with the older, more familiar technologies that have been integrated into ways of life. Safety is not the only concern; autonomy among many other ideas is also important. This is true for the mountain climber, the astronaut, the soldier; it is true for society as a whole as well.

PUBLIC/PRIVATE

Commentators on social regulation often refer to a spectrum or continuum stretching from 'private' to 'public' risks. Thus:

> The risks of overeating are at the most private end of the spectrum, the risks of contaminates in mass-produced food toward the public end. . . . Still more 'public' are risks of diffuse pollution. . . . Finally, there are risks of a global character – the 'greenhouse effect' . . . the risks of nuclear war. These are the most public risks of all – universally shared and individually inescapable.
>
> (Huber 1985: 278)

It is worth emphasizing that many of the approaches people take to evaluating risks involve what Strawson has called *reactive* attitudes, feelings that are appropriate responses to actions within a specific context of interpersonal relationships (Strawson 1974). Thus, football players on the field expect to be smashed by members of the opposing team; in that context, they would hardly resent the risks associated with the brutality of the game. In another context – say, an encounter between strangers on the street – the same behaviour would produce resentment and, possibly, a law suit. The way one appropriately reacts or responds to risk depends less, possibly, on its magnitude than upon the circumstances or 'frame' in which it arises.

Strawson characterizes 'participant reactive attitudes' as 'essentially natural human reactions to the good will or indifference of others towards us, as displayed by *their* attitudes and actions' (1974: 10). People generally have at their disposal a subtle and extensive range of reactive attitudes to bring to bear on private risks. We know when to be angry, when to forgive; we know what kind of further information would matter in determining our response. With public risks – the hazards imposed by large-scale, new and complex technologies – our situation is different. We find it difficult to find a framework of personal feelings, relationships and expectations against which an attitude of anger or forgiveness, for example, would be appropriate.

Private and public risks arise in such entirely different social frameworks – the interpersonal relationships they assume are so incomparable – we should not expect our attitudes towards them to have much in common. A person who endangers your life, for example, by sideswiping you as you drive, will elicit your anger. But you will soften your feelings if you are told that he was pushed by another car, or had to swerve to avoid a pedestrian. Excuses of this sort invite us to separate the risky action from the agent and to direct our ire at the event itself rather than at the person who is the immediate cause of it.

When we consider public risks, we cast about to locate the people who are responsible; sometimes these are hard to identify. We may come up with the name of a corporation, an industrial association or simply the empty pronoun 'them'. What are *they* trying to do to us? Do *they* care about the effects of nuclear wastes, recombinant microbes or whatever over the long run? To characterize relationships in this totally impersonal way is to invite suspicion. And there seems to be little basis in settled expectations and cultural history to suggest how 'they' ought to behave to people who may have to pay the social costs of large-scale technologies.

Commentators sometimes argue that we invest too much from the point of view of safety in reducing public rather than private risks. They point out that we could get more safety per dollar invested by attending less to risks people impose on one another and more to risks they impose on themselves. I imagine this is true. Accordingly, if risk-minimization were our primary goal, or if utilitarian goals like safety and health were our primary considerations, then we should want government to be more paternalistic than it is. We should be less concerned to regulate the risks others impose on us than the risks we impose on ourselves.

Our political theory, however, tends not to be utilitarian. Rather, it regards individuals as separate and inviolable in some sense that does not allow government simply to maximize personal welfare in the aggregate. We do not regard persons, in other words, merely as channels or locations at which welfare is to be found. Rather, we respect the boundaries which separate one individual from another and we require policy to treat persons as ends in themselves and not simply as means to ends.

The central question in social policy with respect to risk is not how we should maximize some quantity, such as efficiency, consumer surplus or safety. The question in the public as in the private law of nuisance has to do with the moral duty to respect the boundaries which separate one person from another. The problem is to determine 'where the harm or risk to one is greater than he ought to be required to bear under the circumstances'.[4] Accordingly, invocations of the efficiency of curtailing private rather than public risk, albeit well argued on the facts, may be beside the point.

We may be led to the question of efficiency, however, simply because we do not have a better vocabulary or a better approach to understanding the impersonal relationships that dominate the social management of risk. Situations of danger and injury call up some of the most anguished and intimate emotions; these feelings often seem appropriate, while intoning the mournful numbers of risk assessment may seem stony or cruel. Much of our unease with current theories of risk management, then, may arise from their apparent detachment from the circumstances of actual harm. We should look a little more closely at the difference between the detached scientific attitude of risk management and the engaged concerns of citizens who confront the possibility of suffering a specific harm.

ENGAGED/DETACHED

In *The Methods of Ethics*, Henry Sidgwick uses the example of a competitive game to point out an interesting distinction in moral psychology. A player as he or she enters such a game may do so simply from the desire to play it; in agreeing to play, he or she will not seek victory as much as 'the pleasant struggle for it' (Sidgwick 1877: 135). As soon as the game begins, this changes. Once the player is engaged in the sport 'a transient desire to win the game is generally indispensable'.

The contrast Sidgwick draws here between our engaged and detached attitudes in playing a sport like tennis applies as well to social attitudes concerning dangers to safety and health. We willingly spend enormous sums in heroic efforts to save people trapped in mines, for example, or to rescue children who may be afflicted by disease. In doing this, we may recognize that we could save many more lives, from a statistical point of view, if we let these miners, children or whomever perish and spent the funds on public health measures or other preventive actions.

When we engage in rescue attempts, however, we have in mind the plight of particular individuals, such as a specific child. Compassion leads us to compare that child with our own; the life we save, in an empathetic way, is our own. We would have little patience, then, for the instruction that the resources we expend might be used more 'effectively' or more 'efficiently' if invested, say, in public hygiene. When our emotions are engaged, we simply *see* what the immediate situation calls for – for instance, that the lost child must be found 'at all costs'.

The context changes dramatically when we consider not isolated dramatic situations but general policies, for public safety, say, and health. In the detached framework of policy analysis, we are morally bound to consider opposing interests, limitations in resources and conflicting intuitions. We may now concern ourselves not with particular but with statistical lives; we think not in terms of the hideous face of Death but in terms of the reasonable upper-bound of Dillingham estimates of the value of life. And thus we may make reasonable policy decisions that result in fatalities which, were we confronted with them, we should go to spectacular lengths to prevent.

It would be easy to criticize this divided consciousness as hypocritical. Yet this kind of hypocrisy, if that is what it is, makes acceptable social behaviour possible. A tennis player ought to take a detached attitude – win or lose, it only matters how you play the game – when he or she agrees to a match. But once engaged, the player should do his/her best to win. Likewise, we have to take a detached perspective in policy to be responsible; but responsibility may demand something else again in a dramatic situation. Our most detached speculations become meaningful for us, however, only in so far as they ultimately engage our sympathetic and compassionate emotions. Just as trying to win at tennis, then, seems essential to the proper play of the game, so trying to save lives must be the goal of regulatory policy.

NATURAL/ARTIFICIAL

In 1983, Bruce Ames published a carefully documented article arguing that the interaction of dietary carcinogens and anti-carcinogens presents a most promising field for cancer research (Ames 1983). Ames argued that many substances which tend to oxidize cells occur naturally in many foods – perhaps most foods – and are far more carcinogenic than many chemical additives, including several which are banned as carcinogens. He also offered evidence to show that natural foods which contain oxidizing and therefore carcinogenic elements also contain elements which are anti-carcinogenic, in so far as they protect cells from oxidation.

The policy implications of Ames's research – even though Ames himself does not necessarily state them – are clear. First, regulatory policy ought to be less concerned with artificial or synthetic chemicals, and more concerned with naturally occurring carcinogens, if it intends to control the primary causes of environmentally induced cancer.

Second, cancer may be prevented by dietary means. Accordingly, the public should worry less about what industry puts in the air, food and water, and more about naturally occurring dietary carcinogens and anti-carcinogens. By persisting in its preoccupation with industrial and synthetic chemicals, according to Ames and many others, regulatory policy swallows a camel while straining at a gnat.

I do not know whether the evidence Ames offers justifies his conclusion concerning the importance of naturally occurring dietary carcinogens and anti-carcinogens. This question may always remain controversial. Suppose, however, the evidence is there. Should we then accept the implications for regulatory policy?

One might look to political theory and to culture, once again, for reasons to think that the government should regulate 'artificial' more strictly than 'natural' risks, even if they are no more dangerous. Three reasons come to mind. First, artificially induced risks, like hazardous pollutants, involve harms for which people are responsible and which they impose on one another. It is precisely this kind of harm – rather than those risks all flesh is heir to – which evokes public resentment and which governments are therefore set up to regulate.

Second, I mentioned that autonomy involves a person's control over the conditions under which they choose and pursue their ends, rather than the degree to which they achieve them. Autonomy is threatened when these conditions come under the control of others. Thus, we do not complain that laws of nature, like gravity, illegitimately thwart our will, even though we are absolutely subject to those laws, and they prevent us from doing many things. As long as constraints are determined by nature, we do not, as a rule, begrudge them as intrusions upon autonomy.

The same is true, I believe, about sources of harm. If the cause of a death is natural, such as lightning, it is still painful, but we may mourn without resentment. With artificially induced risks it is different. Here we believe that someone is at fault, so we are concerned not simply with safety but with responsibility and guilt as well.

Finally, our environmental law, at least according to the National Environmental Policy Act, is not preoccupied with public safety and health. Rather, environmental regulation, most generally conceived, intends to encourage a 'productive and enjoyable harmony between man and his environment' (section 4321).

It is reasonable to suppose that this notion of 'harmony' between man and nature relies upon ideals and ideas familiar in romantic and pastoral imagery. Technological hazards are a problem for environmental policy, therefore, not only because they are hazards but also because they are technological. Their existence recalls a doctrine that had become conventional in romantic literature more than a century ago: 'that of a fundamental opposition of nature to civilization, with the assumption that all virtue, repose, and dignity are on the side of "Nature" – spelled with a capital and referred to as feminine – against the ugliness, squalor, and confusion of civilization, for which the pronoun was simply "it" ' (Miller 1967: 197).

Artists, poets, novelists and others who created romantic and pastoral visions of nature encountered great difficulty in explaining the presence of death in Arcady. The banner '*et in Arcadia ego*', which forms the legend of a famous Poussin painting, expresses the tension between the romantic belief that nature is benign and the scientific evidence that it is not. Bruce Ames's article presents us, once more, with this dilemma of romanticism.

The problem is particularly vexing for Americans because our artists and writers – Whitman, Thoreau, Melville, Hawthorne – recognized that nature, whether represented by the forest or the sea, is often violent, indifferent and inimical to man. Yet our national cultural and religious history typically represents nature as expressive of God's beneficence to and love for man.

The challenge of natural vs technological risk is not to be understood wholly in economic terms or even in terms of autonomy, freedom and justice. Rather, it seems primarily to be a challenge to religion, or at least Protestantism: 'the awful task of re-examining, with severest self-criticism, the course on which it so blithely embarked a century ago, when it dallied with the sublime and failed to consider the sinister dynamic of Nature' (Miller 1967: 207).

FAIR/UNFAIR

When a person takes a risk and loses, we cannot say that he or she consents to that loss, but we may, in some circumstances, say that it is fair that he or she bears it. If a person buys a ticket in a fair lottery, knowing the odds, for example, he or she should gracefully accept the outcome. But if an accident at the chemical plant down the street injures him or her, he or she has a right to claim compensation, even though, by living in that place, he or she may have gambled and lost.

One might try to distinguish 'fair' from 'unfair' risks by appealing to the notion of what it is reasonable for a person to do. This approach would not clarify matters, however, because the notion of reasonableness is so vague. If we define 'reasonable' in terms of 'rational' in a technical sense, for example, then the gamble presented by state lotteries would hardly be reasonable. Even if these lotteries are entirely fair (i.e. they are not 'fixed') the odds of winning are so slim that they do not offer gambles a rational individual would take. The risk one takes by living near a chemical plant, on the contrary, might be perfectly rational. Yet it would be unfair to expect the victims to bear the losses inflicted on them by an unlikely explosion.

I think that the distinction between 'fair' and 'unfair', like that between 'voluntary' and 'involuntary', may be too general and embody too many other concerns to be usefully applied in risk management. Yet we may say that the question whether a risk is fair often depends on whether the context of risk taking is socially beneficial in general. One might ask, for example, if the institutional framework in question serves a socially useful cause. A state-run lottery which funds social programmes, let us say, serves a good cause. We might say, then, that as long as the lottery is run fairly, the gambles are fair, in spite of the absurdly long odds against winning.

Some economic theorists define 'social welfare' in terms of the outcome of perfectly competitive free markets – that is, market exchanges in which all parties understand the terms of the transactions that affect them and voluntarily make those transactions. The hideous results of *laissez-faire* markets of the last century would maximize 'welfare' in this tautologous sense, since workers who went down into the mines (or sent their children), who died railroading or ruined their health in textile mills knew full well the horrors that awaited them. They took gambles and lost. Having defined social welfare in terms of the results of competitive markets of this sort, one may argue that these poor souls should accept deaths and injury graciously in the knowledge that, even if every worker is killed and every child maimed and stunted for life, social welfare as a whole has been maximized. That sort of conclusion would seem to follow from a theory that defines welfare in terms of the functioning of a perfectly competitive market with no externalities.

If we choose to define 'welfare' in a substantive sense the way utilitarians like Bentham and Mill did, in terms of human well-being and happiness, then we will probably think that *laissez-faire* markets, however perfectly competitive, do more harm than good. We should then say that the risks workers took a century ago were unfair, indeed, unconscionable. The reason is not simply that the odds against the workers were so long; after all, the odds of winning in the state lottery are even more ridiculous. One can argue that a state lottery serves (as a form of voluntary taxation) a reasonable purpose, to provide care for the elderly, say, while *laissez-faire* markets, even if perfectly competitive, seem to do very little good and a great deal of mischief if one judges them in a way that makes moral sense.

Whether a particular technological risk is acceptable today may depend on what we think in general of the direction of a technology-based consumer-driven industrial economy. One may not be able to escape arguing over what may be the most general and basic question, namely, the degree to which technological progress actually serves the human good and the extent to which it simply speeds up the rat race.

Americans have typically greeted technological civilization in an entirely schizophrenic way. Americans describe the 'rat race' with literary, cultural, ethical and aesthetic revulsion. And yet Americans have had little difficulty in overcoming this revulsion in order to promote economic growth. Whether technological risks are 'acceptable' or 'fair' may depend more on what we think about *technology* than what we think about *risk*.

Technological risks present essentially the same dilemma today that Thoreau, Emerson, Melville and others described 150 years ago: the problem of reconciling shared ideals and aspirations concerning the quality of our lives with the wants and preferences expanding consumer markets create and satisfy. It is the perennial problem of whether wealth buys happiness – in other words, whether the technology-based economy that leads us to have and to satisfy continually expanding desires leaves us better off in any substantively or morally meaningful sense.

We tend to address this question today not only in literary and artistic but also in legal, economic and philosophical terms. Thus we are prone to speak today of a 'trade-off' between equity and efficiency. Two or even three centuries ago, we might have stated the problem differently. We should have said that we have declined from our ideals to pursue 'the things of this world' or, more precisely, that things are in the saddle and ride mankind.

I do not know how to account for this cultural transformation of literary and artistic themes into legal and philosophical problems. If I understood the reasons for it, I should have begun this chapter by explaining them. Others may understand this change better than I, and I invite them to comment on why it has taken place.

CONCLUSION

Earlier in this chapter, I mentioned that Chauncey Starr argued that we need a double standard for regulating 'involuntary' as opposed to 'voluntary' risk. Starr suggested that an involuntary risk, to be acceptable, must confer benefits 1,000 times greater than a voluntary one of the same magnitude. Commentators following Starr have generally agreed that we need to apply a 'double standard' in regulating risks of equal magnitude, that we should require that greater or lesser benefits be associated with them, if they arise in a different moral setting. Thus, in regulating risks, it is not just the *magnitude* but also the social and moral *meaning* of risk that counts.

I have argued in this chapter that many distinctions beside the 'voluntary/involuntary' pair demand special attention in the management of risk. During the past several years, psychological research has revealed many such distinctions, including several I have not mentioned, each of which seems to require its own 'double standard' if risk-benefit analysis is to make moral sense.

Paul Slovic and Baruch Fischoff, for example, who are leading researchers in this area, have attempted to aid policy makers 'by examining the opinions that people express when they are asked, in a variety of ways, to evaluate hazardous activities and technologies' (Slovic *et al.* 1980: 211). They have found that all sorts of moral factors lead people to object vehemently to one risky activity while cheerfully accepting another – even if the magnitude of the risks and of the associated benefits, from a detached perspective, would appear to be the same.

These moral qualities include: the immediacy of the outcome; the control the individual exercises over it; the extent of uncertainty; the chronic or catastrophic nature of the outcome; the novelty or unfamiliarity of the risk; and the dreaded nature of the result, such as whether it involves cancer. These distinctions, along with the ones I have described, offer only the beginning of what must be a much longer list. Thus, we might conclude that we need to introduce as many 'double standards' into risk management or risk-benefit thinking as there are cultural and aesthetic attitudes towards risk.

A more practical conclusion might be to manage risk by managing technology rather than the other way round. To do this, we should have to think as political communities about the moral desirability of particular uses of particular technologies. Once we determine on moral or cultural grounds that a technology serves a substantive social need, we may then attend to weighing the risks it may involve. To try to abstract the risks away from the technology and deal with them separately is to bring up all the distinctions I have mentioned and many more besides. And these distinctions drag in through the back door of risk management the very problems of technology assessment that perhaps managers sought to avoid.

The proper approach to the regulation of risk, then, may have less to do with *risk* than is commonly supposed. It surely has less to do with the *magnitude* of risk; it has more to do with the *meaning* of risk; and it has the most to do with the way a particular technology fits in with the needs and aspirations of a society or a culture. This suggests that we may be putting too much emphasis on the objectivity of risk assessment. Our most basic and important problem lies in getting the values of risk management right.

NOTES

1 National Research Council, US National Academy of Sciences (1983).
2 Starr comments that 'the indications are that the public is willing to accept "voluntary" risks roughly one thousand times greater than "involuntary" risks' (Starr 1969: 1237).
3 See Thomas Hobbes (1960: especially chapter 13).
4 *Restatement of Torts*, section 822, comment *j*.

16 Thinking, believing and persuading: Some issues for environmental activists

Rosemary J. Stevenson

Some people firmly believe that the hole in the ozone layer is so serious that all life on Earth will cease some time in the next century. Other people believe just as firmly that the presence of a hole is reversible and that all the proposed negative consequences will be alleviated before the end of this century. Yet others believe that there is a hole in the ozone layer and that the people in power are keeping it a secret, and still others believe that there is no hole in the ozone layer at all and that the people who say that there is are scaremongers. What is it that makes different people hold such radically different beliefs? One reason might be a difference in knowledge, but this is unlikely to be the whole story, as we shall see in this chapter.

From not knowing how people reach decisions and formulate conclusions, it is only a short step to making bad decisions and formulating faulty conclusions. Since the current environmental debate has such potentially far-reaching and serious consequences, it behoves us to find out about the way that people do these things. This chapter is written with this aim in mind. In it I discuss how people estimate probabilities, evaluate arguments and draw conclusions both on their own and when they are in groups.

ESTIMATING LIKELIHOODS

How easy is it to estimate the likelihood of different outcomes? For example, given present practices, what is the likelihood that the Earth's natural resources will be depleted during the next century? What is the likelihood that alternative resources will be found? It turns out that people are not very good at making these kinds of estimates. Slovic *et al.* (1982) have reviewed a set of psychological experiments comparing the assessment of risk with actual frequency data from the past for a wide variety of risks ranging from crime and disease to motoring accidents. They found that people show large errors of judgement; for example, people radically overestimate the risk of dying from accidents as compared with illnesses. But whether a person underestimates or overestimates depends (among other things) on the availability of information and its vividness.

Availability of information in this context refers to the information that a person thinks about when making an estimate, rather than the sum total of information that is known about the topic. Its importance in making inferences about probabilities has been highlighted by Tversky and Kahneman in a series of influential papers dating from the early 1970s to the present time. They have shown that intuitive assessments of probability bear scant relation to actuality and are governed instead by the use of a small number of rules of thumb which often lead to systematic errors and biases. One of these rules of thumb is availability.

Availability

Tversky and Kahneman's original use of the term 'availability' in 1973 was quite specific and limited to likelihood judgements. They proposed that 'A person is said to employ . . . availability . . . whenever he estimates frequency or probability by the ease with which instances or associations can be brought to mind.' On the face of it, this sounds like quite a reasonable way of estimating likelihood and one which should lead to correct answers. For example, a doctor examining a patient with suspected food poisoning might recall a number of case histories of patients presenting similar symptoms in the doctor's past experience, and use this information to estimate the likelihood of rival diagnoses in the present case. Provided that the doctor's memory is accurate and that her experience encompasses a sufficiently large and unbiased sample of examples, this should provide a fairly accurate guide to probability. Unfortunately, it doesn't.

One of the main reasons that it doesn't is because human memory does not work in such a machine-like way. People are more likely to remember vivid events than mundane ones, so mundane events will not be retrieved because they are less likely to have been stored in memory in the first place. Even when events have been stored in memory, some are easier to recall than others. Availability biases can also be triggered by expectancies and prior beliefs. For example, when clinicians are provided with a series of case descriptions they will perceive patterns of relations between clinical tests and diagnoses which conform with their prior beliefs, even though the relationships are not actually present in the data (Chapman and Chapman 1967). People are more likely to store in memory and to recall information which favours a prior belief. Even when recall is accurate, the information that has been stored in memory may be biased. I have already mentioned that people radically overestimate the risk of dying from an accident compared to dying from an illness. The availability explanation of this is that the media provide a highly selective coverage of violent and spectacular accidents, such as aircraft crashes, but they provide very little coverage of deaths by routine causes such as strokes and heart attacks unless the person is famous. Certainly media coverage of environmental risks is selective, and there is little relation between amount of coverage and actual risk (Greenberg *et al.* 1989).

These characteristics of human memory doubtless affect our ability to estimate environmental risks and may well account for some of the differences in views which people hold on these issues. However, Evans (1989) points out that the Kahneman and Tversky paper is important not only in terms of its original objective of explaining a major cause of bias in likelihood judgements but also for wider reasons. The kinds of biases in memory retrieval that they identify probably influence tasks other than likelihood judgements. Many other kinds of judgements can also be affected by information that we 'call to mind'. For example, a public tribunal on the siting of a nuclear waste dump needs to weigh the evidence for and against the proposal. If people's memories are affected by availability then the eventual judgement could be biased accordingly.

When experts try to assess the probability of an event occurring (let us say an engineer wants to predict the probability of a catastrophe in a nuclear power plant) they often use tree diagrams as a way of trying to ensure that they have considered all of the possibilities. By breaking down a problem in this way, we can obtain a more accurate probability estimate. Of course, there is no guarantee that all the possibilities are included in the decision tree, particularly when some of the possible hazards may not yet be known. So people include a category of 'all other problems'. It has been found though that even experts underestimate the likelihood of risks due to these 'other problems' (Fischoff *et al.* 1978). When something is not represented in our analysis, we tend not to think of it. Thus, while frightening news stories, such as those about Three Mile Island and Chernobyl, tend to lead to overestimates of both their frequency (though they occur very rarely) and their risks, failure to consider what information might be omitted from the analysis may lead to underestimates (Baron 1988).

Vividness

In a study of environmental risk prediction, Baird (1986) asked students to estimate the perceived risk over time for nine environmental factors: population, food, water, environmental pollution, disease, genetic defect, mental disorder, nuclear power and nuclear war. In general, students were most optimistic about disease factors and most pessimistic about nuclear power and nuclear war. However, the point I wish to make is that the students' estimates of future risk did not vary whether they were asked to think about relatively short time scales (20 years), intermediate time scales (2,000 years) or very long time scales (5,000 years). Where people's judgements are concerned, the future does not have a great deal of impact. This may be due to its lack of vividness.

The concept of vividness as a cause of the selective use of information has been discussed by Nisbett and Ross (1980). They proposed the notion to explain why people differentially weight the evidence when making social judgements. For example, court-room juries may be heavily influenced by the dramatic, though notoriously unreliable, evidence of eyewitnesses and relatively unimpressed by the dull meticulous testimony of expert forensic scientists. Similarly someone may be much more influenced by a colleague's racy story about corporate greed than by a scientist's report of the difficulties of depositing chemical waste.

In order to explain such apparent biases, Nisbett and Ross propose that people overemphasize vivid, concrete information and underemphasize dull, pallid and abstract information. Specifically, they suggest that vividness of evidence is determined by emotional interest, concreteness and things in the here-and-now rather than things that are distant and in the future. As Evans (1989) points out, it is no coincidence that these factors are also associated with things that are easy to remember.

EVALUATING ARGUMENTS AND DRAWING CONCLUSIONS

Economic growth increases the hole in the ozone layer.

The hole in the ozone layer is a bad thing.

Therefore economic growth is a bad thing.

This argument looks plausible, given the premisses that have been stated. But many people would try to argue that it is not. Of course, many people would argue that the first premiss is false, but that is not the point I wish to make. The point I wish to make is that people do not evaluate arguments in a vacuum but in the context of a set of beliefs and values that they have about the world and how it is (or should be) organized. This holds for drawing conclusions as well as for evaluating them, but in the former case there is the additional problem that people seem to be loath to seek evidence which disconfirms their conclusion. We will see later that this is not a deliberate ploy to see the world from our own point of view but an aspect of cognitive processing that is hard to carry out. First, though, I will look at what has come to be called in the cognitive psychological literature 'the belief bias effect'.

The belief bias effect

People persist in their beliefs frequently in the face of evidence to the contrary. Belief persistence in itself is not necessarily a bad thing. Michael Faraday persisted in believing that electrical currents could be induced with magnets despite several failures to produce such currents experimentally, and finally of course he succeeded (Tweney 1985). Nevertheless, there is a long history in the psychology of thinking which shows that people are loath to accept logical conclusions of arguments if the conclusions violate their beliefs and values. Conversely, people are more likely wrongly to accept as valid an illogical conclusion, if it coincides with their beliefs. One of the earliest reported studies is that of Wilkins (1928) and since then the work has suggested not only that conclusions are sieved through a net of personal beliefs but also that people's reasoning ability may be impaired if the potential conclusions run counter to their beliefs (Oakhill *et al.* 1989).

Belief persistence involves two very similar types of biases. One is that we weight evidence in favour of a belief and neglect evidence against it, and the other is the failure to search impartially for evidence. For example, if I believe that the water supply is contaminated I will pay more attention to reports of waste disposal into the rivers than to reports of an industry's attempts to recycle waste.

How can we tell whether people are weighing all of the evidence? One way is to make the assumption that neutral evidence should not strengthen a belief. By 'neutral evidence', I mean evidence that is, on the whole, equally consistent with a belief and its converse. Let us call this the neutral evidence principle. Lord *et al.* (1979) showed an apparent violation of the neutral evidence principle in a situation where such errors in daily life have deadly serious consequences. Their neutral evidence consisted of mixed evidence, that is, there was some evidence in favour of the belief and equal evidence against it. They first selected people who had indicated by means of a questionnaire that they favoured or were opposed to capital punishment. Then the people were asked to read two reports, one presenting evidence that capital punishment was effective in reducing murder rates, the other presenting evidence that it was ineffective. Both reports were presented as if they were journal articles, but were in fact constructed by the experimenters. Lord *et al.* found that the effect of each report on a person's belief was stronger when the report agreed with the belief than when it did not. People found the report that agreed with their belief 'more convincing', and they found flaws more easily in the report that went against their belief. In the end, the people became stronger in their initial belief whatever its direction.

This study is disturbing, because it suggests that evidence is useless in settling controversial social questions. Of course, the results may be limited to certain types of cases and to the demands of the particular study. Nevertheless, the study suggests that a major mechanism of the persistence of beliefs involves distortion of one's perception of what the evidence would mean to an unbiased observer.

What determines belief persistence in the face of evidence to the contrary? There seem to be a number of factors, including wishful thinking, what people count as good thinking and also selective exposure to information.

Wishful thinking has been observed in people's reluctance to accept unpalatable conclusions, and in their enthusiastic acceptance of desirable conclusions. McGuire (1960) gave people questionnaires containing a number of propositions such as the following:

> Any form of recreation that constitutes a serious health menace will be outlawed by the city health authority.

> The increasing water pollution in this area has made swimming at the local beaches a serious health menace.

> Swimming at the local beaches will be outlawed by the city health authority.

The more unpalatable people found the conclusion, the more unwilling they were to accept it, despite the argument's impeccable logic. It is interesting to notice this environmental theme occurring in a study conducted thirty years ago. We would perhaps expect the desirability of such a conclusion to have changed over the years, an expectation which might cause us to ponder the range of influences on our thinking.

Most people wish to regard themselves as good thinkers, but may not have a clear idea of what good thinking is. For example, our 'role models' in this regard are usually experts in their fields. Because experts know the answers to most of the questions that the rest of us are able to think about, they do not usually have to consider evidence or counterevidence. If we admire experts, then we may come to admire people who appear knowledgeable and decisive rather than thoughtful and evaluative. We are thus encouraged to know rather than to think. Of course, knowledge is crucial to develop our thinking, but knowledge itself is not thinking.

People also maintain their beliefs by exposing themselves to information that they know beforehand is likely to support what they already want to believe. Socialists tend to read socialist newspapers, and conservatives tend to read conservative newspapers. During the 1964 election campaign in America, people were given an opportunity to order free brochures either supporting the candidate they favoured or supporting that candidate's opponent. They were given the opportunity to sample the brochures first. When the arguments in the sample were strong and difficult to refute, people ordered more brochures supporting their own side than brochures supporting the other side. When the arguments in the sample were weak and easy to refute, people tended to order more brochures supporting the other side (Lowin 1967). People can strengthen their own beliefs by convincing themselves that the arguments on the other side are weak or that their opponents are foolish, as well as by listening to their own side.

If people seek out belief confirming information when evaluating arguments and conclusions, what can we say about situations where people are trying to form a conclusion? Here the desire for confirming evidence to support a hypothesis is equally strong and has come to be called the confirmation bias.

Confirmation bias

Confirmation bias is perhaps the best known and most widely accepted notion of inferential error to have come out of research into human thinking. People have a fundamental tendency to seek information consistent with their current beliefs, theories or hypotheses and to avoid the collection of potentially falsifying evidence. Numerous psychological studies have shown that people fail to discover general rules when required actively to seek relevant evidence. This is because they adopt strategies designed to confirm rather than refute their hypotheses. It is this bias which is blamed for the maintenance of prejudice and irrational beliefs. In a different context, it has also been argued that the bias operates directly against the dictates of Popper and other philosophers of science concerning the correct means of conducting scientific enquiry. According to Popper, scientists should try to falsify their theories and to examine alternative hypotheses wherever possible, whereas intelligent people in laboratory simulations of scientific reasoning generally appear to do precisely the reverse (e.g. Wason 1960).

The general view which prevails in the literature is that this is a kind of motivational defect. For whatever reason – vanity, maintenance of belief structures, fear, envy – people are assumed to be actively attempting to verify rather than falsify their hypotheses. But I subscribe to a different view, one that has been most explicitly stated by Evans (1989). He reviews the evidence on confirmation bias and comments that some studies have not shown such a bias. After a careful examination of those studies that do and those that do not reveal the bias, he argues that people find falsification difficult. However, when people do encounter falsifying evidence they generally do reject their initial hypotheses. And when people are specifically instructed to test negative predictions they do so, although general instructions to try to disconfirm their hypotheses are not successful. From this pattern of findings Evans argues that the confirmation bias is not a motivational problem; people will try to falsify when instructed to do so. But people do not find it easy to know *how* to falsify; general instructions just to do it do not prove helpful. Evans suggests therefore that the confirmation bias is a feature of our cognitive processes and it is hard to overcome because it requires a conscious effort as well as an understanding of how to go about doing it.

Can we be more explicit than this about the pervasiveness of the confirmation bias? I think we can. Many of our cognitive processes are rapid, automatic and unconscious. We would not be able to make our way about the world if we had to deliberate on every thought and action. Our use of language is an example of this. We write, speak, read and listen with great facility, drawing upon immense reserves of knowledge of the topic under discussion in order to do so. When we listen to someone else, for example, part of what it means to understand what they say is that we are able to construct mentally a model of the world that is being described. We do this automatically, without conscious thought. Once we have constructed a model of this world that is being described, we can draw inferences from the model. If I tell you that I have a pet bird, you may readily infer (among other things) that it may sometimes fly around my living room. And yet you did not need to deliberate to make this inference. Indeed, if you had consciously deliberated over it, you may well have been able to falsify the inference, not simply because I may never let it out of its cage but because it might be a non-flying bird, such as a penguin, unlikely though this may be.

It seems that people readily construct mental models of what they read or hear and effortlessly derive inferences from them. But what people find hard is to try consciously and deliberately to make the model false. Johnson-Laird (1983) argues that it is this ability to make a conscious, deliberate attempt to falsify such mental models that makes people rational. But this ability requires conscious effort and attention and people find it difficult. The relative ease with which we can construct a mental model based on what we already know doubtless accounts for the biases in thinking that I have described. The remedy would seem to be greater deliberation, and yet it is our non-reflective model building skills that enable us to act in the world with such apparent ease. If we deliberated over everything, we would do nothing.

If individuals are so fickle in their ability to evaluate opposing points of view, perhaps we should consider group decisions. After all, the ordinary person in the street has little power over the kinds of policy decisions that are needed to solve the problems of the environment. If concerned individuals can convince policy makers to take the issue seriously, will we fare any better? Is the thinking of the group more than the sum of its individual parts? It seems that it is, but not necessarily for the better.

GROUP DECISION MAKING

When people are in groups a number of things can happen in addition to the biases that I have described above. I will mention two of them: the tendency to make extreme judgements and the tendency to collude with the leader.

Extreme judgements

In the 1950s, the social critic William H. Whyte (1956) claimed that groups, notably those within business and governmental organizations, tended to make safe, compromise decisions instead of risky, extreme decisions. Whyte assumed that this tendency explained why organizations failed to be as creative and innovative as individuals. But research in the 1970s has shown that groups tend to make more extreme decisions than their members would make as individuals. These decisions may be more risky but they may also be more cautious. In this kind of research people are presented with a moral dilemma and each individual is asked to make a decision about it before joining the group. The group then has to form a joint decision about the dilemma. It is this joint decision that is more extreme than the individual decisions. For example, when a group of high-school students, who were either high or low in racial prejudice, discussed racial issues, those who were low in prejudice became less prejudiced and those who were high in prejudice became more prejudiced (Myers and Bishop 1970).

It seems likely that two factors contribute to this group polarization. One is that group members who initially hold a moderate position about an issue will move in the direction of the most persuasive arguments, and this will eventually move the group to a more risky or cautious decision. The other is that people perhaps view it as socially desirable to agree with the average tendency of the group members. When this happens, a group member may adopt a slightly more extreme position than the group average, again causing the group as a whole to move towards a more risky or cautious decision.

Group collusion

There is a tendency for small, cohesive groups to place unanimity ahead of critical thinking in making decisions. Janis (1983) has called this phenomenon *groupthink*. It is promoted by several factors: the desire to maintain group harmony, a charismatic leader, feelings of invulnerability, discrediting of contrary evidence, fear of criticism for disagreeing, isolation from outside influences and disparaging outsiders as incompetent. Groupthink may well be responsible for some of the more disastrous decisions that are made by our politicians. Janis discusses two decisions by J. F. Kennedy to illustrate groupthink and how to overcome it. Both decisions involve Cuba.

On 17 April 1961, 1,400 Cuban exiles who had been trained by the CIA landed at the Bay of Pigs in an attempt to overthrow Fidel Castro. Within three days about two hundred of the invaders had been killed and the rest captured. According to Janis, President Kennedy and his advisers had succumbed to groupthink. Kennedy, a strong leader, played an active role in group discussions; the group failed to seek outside opinions; and they ignored contrary information about the expected strength of Cuban resistance to the invasion. While it was clear to the group that many risks were involved, little serious consideration was given to the question of what would happen if the worst outcomes materialized – as they did, in practically every case.

This particular story ends on an optimistic note, since Kennedy and his colleagues appeared to learn from their mistake. During the Cuban missile crisis of October 1962, practically the same group of people avoided groupthink in reaching a decision on how to convince the Soviet Union to remove its nuclear missile bases from Cuba. This was a monumental decision, because it was the closest that the United States has ever come to nuclear war. Kennedy permitted his advisers to hold some meetings without him. His advisers invited contributions from outside experts. And members of the advisory group were encouraged to play 'devils's advocate' by countering all proposals. Kennedy and his advisers decided to assure the Soviet Union that the United States would not invade Cuba in return for the removal of the nuclear missiles. This proposal was accepted and a possible nuclear confrontation was avoided.

INFLUENCE AND PERSUASION

Does the majority always determine the outcome of group decision making? Despite our own desires for omnipotence on the one hand and our suspicions about other people on the other, the general answer is yes. But under certain circumstances, minorities may influence group decisions. This is perhaps where the ordinary person in the street may convince a more powerful majority to make decisions about environmental issues. I will say more about influencing majorities shortly. First, let us find out a little more about how persuasion works.

Persuasive messages may take a central route or a peripheral route. A message that takes a central route relies on clear, explicit arguments concerning the issue at hand. This encourages active consideration of the merits of the arguments. By contrast, a message that takes a peripheral route relies on factors other than the merits of the arguments, such as characteristics of the source or the situational context. The credibility of the source of the message is the most influential factor in the peripheral route and it seems to depend on the source's expertise and trustworthiness. By contrast, characteristics of the message itself may contribute to either the central or the peripheral route. It turns out that it is not always desirable to present arguments that only support a position. When talking to people who already hold a belief, a one-sided argument will be effective. But when talking to people who hold a contrary belief, then a two-sided argument is more persuasive. Two-sided arguments are effective because they enhance the credibility of the source and, as a consequence, decrease counter-arguing by the listeners. They contribute, therefore, to the peripheral route.

But what determines whether the central route or the peripheral route will be more effective? An important factor is the relevance of the message to the listener. When a message has high relevance to the listener, the central route will be more effective. When a message has low relevance to the listener, the peripheral route will be more effective. This was demonstrated in a study of student attitudes towards recommended policy changes at a university that would be instituted either the following year (high relevance) or in ten years (low relevance). Students who were asked to respond to immediate policy changes were more influenced by the quality of the arguments (the central route) than by the expertise of the source (the peripheral route). On the other hand, students who were asked to respond to long-term policy changes were influenced more by the expertise of the source than by the quality of the arguments (Petty *et al.* 1981).

If you are part of a minority and wish to influence group decisions, social psychologists recommend that you follow several well-established principles. First, rational arguments for a position are more effective than emotional arguments. This means taking a central rather than a peripheral route to persuade the majority to take your position seriously.

Second, absolute confidence in a position with no sign of wavering is a crucial ingredient for success. Uncertainty about a position will lead the majority to discount it. Third, the position must be consistent over time. Any inconsistencies will lead opponents to discredit the arguments. Fourth, the quality of patience is a necessity. Though majorities may initially dismiss minority positions, the passage of time may make them ponder the evidence that has been provided and make them gradually change their positions. Fifth, there should be at least one other person arguing the same thing. A minority of two is much more influential than a minority of one (Nemeth 1986).

LIVING WITH INCONSISTENCIES

You may have detected some inconsistencies in the last paragraph with some of the comments that I have made earlier. Let me try to disarm my critics by making them explicit. I have implicitly assumed that environmental concerns are minority concerns. This seems to me to be a reasonable assumption since most national governments pay little more than lip-service to the idea that policies should reflect these concerns. Because of this, I outlined the ways that minorities may influence majorities. One of these ways is by taking the central route to persuasion. But earlier I said that the peripheral route to persuasion was most effective when the issue had low relevance to the listeners, and the example I used of low relevance was a long-term rather than an immediate change.

Yet many of the more serious environmental risks are ones that cast a shadow over the intermediate and long-term future rather than the present or immediate future. This would suggest that a peripheral route to persuasion about environmental concerns is more suitable than the central route. There seem therefore to be arguments in favour of either view: the central and the peripheral route.

I also said that, to influence a majority, absolute confidence in the alternative view is necessary. But this seems to fly in the face of the ideas presented in the first part of the chapter that the key to clear thinking is the ability to consider alternative positions. Of course, persuasion is not the same thing as thinking. But to persuade by means of a single, clear position is only a short step from believing a single, clear position and falling prey to all the biasing effects that I described earlier.

I will not attempt to resolve these conflicts. Indeed, they may not be resolvable. Or if they are, they may require much time and patience to grapple with things that we all find hard to take – like accepting conclusions that we do not find congenial. What I have tried to do is to indicate some of the difficulties that I foresee in trying to come to grips with the consequences of our collective actions in the world. To leave open the inconsistencies and the conflicts is to turn a mirror on the comments with which I began this chapter – comments which highlighted inconsistencies and conflicts between individuals over a single issue. To do something about the environment we need to acknowledge these inconsistences and conflicts, both individually and collectively. We may be better placed to do that if we know the kinds of things that can lead people astray when they think about risk and when they evaluate and derive conclusions.

17 Rethinking resources

Stephen Sterling

For many years, much discussion of resources as a global issue centred on the question of whether the world is 'running out of resources'. This was a major concern at the United Nations Conference on the Human Environment of 1972 which marked the first wave of environmentalism in the modern age. Now, some two decades later, amidst a second and greater green wave, the focus has changed. The finite nature of resources is no less real, but debate surrounds the implications of their 'sustainable use'. At the same time, global pollution levels appear a greater threat than resource depletion. Yet the two are related; pollution is an inevitable consequence of resource use, and also affects the productivity and availability of natural resources.

The realization that such issues are deeply interwoven is becoming increasingly widespread, and the debate has become more sophisticated as the complexity of problems has become apparent. As the seminal Brundtland Report said: 'Compartments have begun to dissolve . . . these are not separate crises: an environmental crisis, a development crisis, an energy crisis. They are all one' (World Commission on Environment and Development [WCED] 1987: 4). Thus resource use cannot be understood in isolation. Scratching beneath the surface of this issue, we immediately find issues relating to politics, economics and the future – and underlying these, questions of ethics, equity and conflicting value systems.

In the environmental climate of the 1990s, we are exhorted to remember the five Rs in relation to resources: reuse, repair, recycling, return (recyclable materials) and refuse (unnecessary packaging and goods). They are indicative of a further and more fundamental 'R' – the rethinking which is beginning to take hold across societies as people realize the implications and urgency of the environment–development crisis. The concern that for years was the province of environmentalists is now taken seriously by public, governments and institutions worldwide. The argument centres on how sustainable societies on a sustained Earth may be achieved, and a critical part of the debate is our whole approach to resources and their use.

KEY ISSUES

From a global perspective, the fundamental issue is the carrying capacity of the Earth, which is determined by the interrelation of population size, resources and environmental systems. Whilst this capacity cannot easily be quantified, it is clear that current patterns of population growth and resource use cannot be sustained. A number of conflicting variables are involved: a finite and declining amount of available resources on Earth, an increasing world population and rising material aspirations in both the rich and the less developed countries. At the same time, the global economic system overall encourages wasteful resource exploitation rather than resource conservation. In addition, degradation of the natural environment from waste products of modern global society is seriously impairing the functioning of ecosystems, thus reducing their ability to cope with that same pollution and therefore also reducing the natural productivity on which the human economy depends. In a sense, notice of environmental limits is being served in the form of such feedback phenomena as global warming, ozone layer depletion, acid deposition and soil loss. Unprecedented demands are being made on the Earth''s resources, at the very time when global pollution indicates that less should be used, or at least that resources must be used very differently. Meanwhile, lack of knowledge about the behaviour and stability of stressed global environmental systems makes fundamental change in resource use patterns all the more urgent.

DEFINING RESOURCES

It may be useful to examine what we mean by resources, as different groups tend to attach different meanings to this word and such meanings hide different value bases. Dictionary definitions often define 'resource' in terms of a supply of aid or support. To economists, a resource is anything that can be used for economic activity including land, minerals, plants, animals, crops, fuels, capital and human labour and skills. The first six groups are classed as natural resources, which are further subdivided into 'non-renewable' and 'renewable'. The former are substances which have evolved over geologic periods of time and therefore effectively cannot be replaced. These include such substances as metal ores, coal, oil and natural gas.

Renewable sources are those which derive from or are powered by solar energy and which are capable of regenerating themselves continually from this input. These include animals and fish, plants and timber, soil, rain, wind and tidal energy. (Strictly speaking solar energy is not renewable but it may be regarded as infinite.)

However, these definitions mask a more complex picture. In particular, renewable resources are only self-generating if they are allowed to be so. Overexploitation effectively converts some renewable resources into non-renewing resources. For example, globally soil is being lost at a rate much faster than it can be replenished, while all species extinctions are of course final. Such destruction represents a loss of both resource 'income' and 'capital'.

PATTERNS OF USE

For most of human history, people's material needs were relatively modest. However, each technological advance from the earliest times increased the demands for materials in absolute terms. Each innovation tended to increase energy use and the throughput of materials, accompanied by the decline of old technologies. The pattern still applies, and although complicated by the emergence of material- and energy-efficient technologies which may locally decrease or stabilize demand, the net trend is greater resource use. Indeed, seen in the time frame of history, the growth of resource use within the last 100 or so years has been dramatic in the extreme. For example, more coal has been used since the mid-1940s than was used in previous human history.

Whereas pre-industrial societies hardly affected the Earth's stock of natural resources in net global terms, modern global society is founded and dependent on using the capital this represents, especially fossil fuels. The whole modern industrial edifice rests almost exclusively on the energy subsidy that has been made available since the exploitation of these fuels. These took some 300 million years to form, but more than half of the world's total reserves have already been consumed.

From one perspective, this has allowed unprecedented prosperity. According to the Worldwatch Institute, 'on average, the additional economic output in each of the last four decades has matched that added from the beginning of civilization until 1950' (Brown 1990b: 3). However, two key problems, of equity and sustainability, are now bearing heavily on conventional assumptions relating to wealth generation.

Use of resources is not equitable. First, some countries are blessed with much greater mineral and fuel reserves than others. Thus the developed countries possess about two-thirds of the world's fossil fuel resources. Second, the per capita consumption of resources varies widely between and within countries. For example, the average American citizen consumes about 300 times as much energy as the average Somalian or Bangladeshi. Taken globally, the developed countries consume about two-thirds of the world's resources, but only have about a fifth of the world's population (Ekins 1986: 13). At the same time, natural resource exports are a key element of the economies of most less developed countries, which 'face enormous economic pressures, both international and domestic, to overexploit their resource base' (WCED 1987: 6). As Trainer points out: 'we could not be so affluent if we were not taking most of the resources and gearing much of the Third World's productive capacity to our own purposes' (Trainer 1989, quoted in Gribbin and Kelly 1989: 77). Quite apart from the implicit ethical issues this pattern raises, the Brundtland Report warned that 'Many present efforts to maintain human progress . . . are simply unsustainable. They draw too heavily on already overdrawn environmental resource accounts to be affordable far into the future without bankrupting those accounts' (WCED 1987: 8).

If productive resources are overexploited the total stock will diminish. With depleted stocks, the productivity of the resource is reduced, and there is a temptation to use up stocks still further – thus perpetuating a downward cycle of positive feedback which if not checked results in a crash. Such a disastrous pattern has been repeated many times at local level – for example, with regard to fish stocks, or the fuelwood situation in some less developed countries where marginalization, poverty and population pressures force local people to over-use supplies despite their knowledge that this use cannot be sustained. From a global perspective, there are indications that some of the world's natural resources are already past peak biological productivity (Worldwatch Institute 1990b). Thus there has been growing emphasis over the last decade, particularly since the publication of the *World Conservation Strategy* (IUCN 1980), on the *sustainable* use of resources. As many writers have pointed out, this means rethinking our economic assumptions.

ECONOMIC ASSUMPTIONS UNDERLYING RESOURCE USE

The problems associated with resource use and depletion are largely rooted in the assumptions that have underlain economic activity in the modern era. The neo-classical economic theory that has largely governed western economic activity for the past century, and has formed the basis of the global economic system, is based on the implicit assumption that the economy is an open system, unconstrained by environmental resource limitations or capacity to absorb waste products. The goal of the system has been to maximize welfare which was based on consumption. The model is fundamentally *linear* and is based on maximization of through-put of energy and materials from the environment and to the environment. As Boulding writes:

> The human race is almost the only living being that has developed a linear economy, moving materials from wells, mines, and soil into products that are then distributed into dumps or flushed down to the oceans or burned in the atmosphere. This is obviously a temporary arrangement.
>
> (Boulding 1978: 296)

Many environmental resources such as air and water are regarded as free, and environmental costs such as pollution regarded as 'external diseconomies'. The costs to the environment in terms of pollution or degradation have not been reflected in price mechanisms. Further, economic policy has not distinguished between renewable and non-renewable resources. Thus, conventional economic theory has had little or no regard for environmental factors. The welfare of future generations has not traditionally been taken into account, as it was assumed they would inherit an even more prosperous world.

Economic wealth has been created by the transformation of natural resources which have been regarded as basically infinite. Such transformation has been seen as necessary to a healthy economy, and the rate of conversion from raw material to economic product has grown not only through increasing populations and rising expectations, but through the artificial stimulation of demand from advertising, built-in obsolescence in manufactured goods and the production of goods peripheral to human needs.

For most of this century, the driving force in the increase in output and consumption of materials has been the belief in the desirability of material progress. Further, the assumption was that development patterns in the affluent North would be emulated by the less developed countries of the South who would in turn share in the new global prosperity.

The assumptions behind this confidence were dealt a major blow in 1972 with the publication of the Club of Rome's *Limits to Growth* study. For the first time, a computer model predicted that continuing economic growth was ultimately not compatible with finite resource levels, and would therefore be constrained by the limits of resource use and the pollution that this generated. The report had a major impact in the environmental debate of the early 1970s, but new technologies of substitution and discovery of new mineral reserves tended to blunt the arguments and message of the study, and its modelling assumptions were criticized. Nearly twenty years on, with a new rise in environmental concern, the limits-to-growth debate is resurfacing but the capacity of ecological systems to withstand environmentally damaging activity and pollution is causing more concern than resource availability. (Supply problems with non-mineral resources in particular are not expected until well into the next century – WCED 1987: 59.)

Whereas we are accustomed to thinking about the effect of economic activity on the environment, we need now to reverse the priorities if we wish to sustain economic activity. As the Brundtland Report noted: 'we are now forced to concern ourselves with the impacts of ecological stress – degradation of soils, water regimes, atmosphere, and forests – upon our economic prospects' (WCED 1987: 5). Neither economic systems nor environmental systems can be considered separately from each other.

The need for integrative thinking is beginning a revolution in economic theory. Whereas conventional economics regards natural resources primarily in terms of the supply of raw material inputs to the production process, a newer green economics is trying to take a much broader view of the entire system, locally and globally. This involves taking account of:

- pollution costs in extraction of materials, in their transport, in the production process, in the disposal of waste products and in the use and final disposal of the product;
- depletion costs where non-renewable resources become scarcer or less available and renewable resources become less productive;
- degradation and rehabilitation costs, where natural systems are harmed to the extent that they are unable to cope with pollution and thus investment in environmental clean-up and rehabilitation is necessitated;
- costs to future generations in terms of depleted resources and pollution.

Such criteria reflect an unprecedented situation globally. Humanity has never before had to think about the total and global effects of local actions, or the effects of present activity on the welfare of future generations. Many writers see the emergence of 'eco-economics' as contributory evidence of a necessary cultural 'paradigm shift' from a mechanistic and reductionist world view towards a systemic and holistic view. The extent of any current cultural shift can perhaps only be judged from the vantage point of some future date, but the model which this argument affords is useful in clarifying the philosophic and attitudinal bases of current resource use, and consequent changes which must be made. The basic paradigm model, as it applies to resource use, is shown in Table 17.1. Current consensus may be thought of as still largely rooted in the conventional paradigm, but attempting to embrace the philosophical and practical implications of the eco-economics view.

RETHINKING THE CONCEPT OF 'RESOURCE'

A rethinking of values and assumptions involves a re-evaluation of the concept of 'resource'. The word immediately defines something in terms of its utility, or potential utility. It implies that the material has no intrinsic value in terms of its being or existence until it is acquired, extracted or used. If something is labelled 'a resource' it is deemed to have instrumental value – usually in terms of producing something of greater economic value. This thinking takes little account of intrinsic values in nature, of aesthetic values or of the value of the multi-functions that a 'resource' may perform whilst left unexploited or at least undamaged – for example, the role that a forest performs in regulating micro- and macro-climate, water flow, soil stabilization and so on. Locally and globally, 'natural resources' are the great life-support systems which

Table 17.1 Value orientations relating to resource use: a model

Conventional economics	'Eco-economics'
Resources are raw materials for transformation	Primary wealth in products and services of a healthy biosphere
Maximum throughput of energy/materials to drive growth	Low throughput to conserve stock
Welfare through maximum consumption and 'trickle down' from growth	Welfare dependent on equitable resource use and low-growth or steady-state economy to meet human needs
Pollution and its effects are externalities	All pollution and social costs must be counted as part of the total economic/ecological system
The economy is an open system, 'separate' from nature	The human and natural economies form one closed system
Resource limits are solvable by technology	Ecological and entropic limits are paramount; technology geared towards sustainability

provide crucial 'free' services including maintaining the quality of the atmosphere, climate and water cycle regulation, generation of soil, disposal of waste and nutrient recycling, maintenance of a vast gene bank and so on. Through the prevalence of a mechanistic view of nature, this role has long been accorded less importance than providing raw materials for economic processes. Ironically, even from this utilitarian view, current patterns of resource use are self-defeating. The continuing loss of biological diversity not only endangers life-support systems but reduces humanity's options in utilizing new sources of genetic breeding stock, medicines, energy and food.

Then, too, the concept of 'resource' does not favour conservation of resources for the benefit of future generations. The significance of this point is perhaps appreciated against the dearth of linguistic alternatives for the word 'resources' which might also convey their intrinsic value. Without a useful alternative, the orientation of our thinking tends to lie with exploitation rather than conservation of natural wealth.

Redefinition is a possibility: 'Resources *are* not: they *become*: they are not static but expand and contract according to human wants and human actions' (Zimmerman 1951: 15). Further, natural resources are not fixed and static but are identified by us as 'resources' at some stage in a wider cycle. For example, the plants we use have depended on air, light, water and soil minerals for their growth. Therefore we should value not only the world's plant resource, but also the conditions which allow it to grow. So these resource-forming and resource-sustaining conditions are also key resources.

It is not only 'raw materials' that are transformed in the economic process but the nature of the living environment too. Instead of focusing on the input of resources into a linear economic process, we must think about the ecology of resources. Thinking in this way marks a shift from a mechanistic and linear view towards a holistic and systemic one which more accurately reflects the dynamic and cyclical processes by which the environment works.

TOWARDS SYSTEMIC THINKING

Among the first thinkers to outline this mode of thinking in economic terms was Kenneth Boulding (Boulding 1966), who pointed out that the conventional 'open system' view of the economy did not accord with the reality of the Earth as a closed system (notwithstanding the constant input of solar energy). Instead of the 'cowboy economy', Boulding stressed that we would in future require a 'spaceman economy' where the Earth is seen 'as a single spaceship, without unlimited reservoirs of anything, either for extraction or for pollution and in which, therefore, man must find his place in a cyclical ecological system' (1966: 77). In the spaceship economy, the throughput of materials would be minimized rather than maximized. Indeed, it would be the maintenance of the stock of resources that would be the primary aim, rather than the flow.

Fundamental to this thinking, and to green economic thinking since, is the inescapability of the laws of thermodynamics. The first is that matter-energy cannot be created or destroyed. Therefore, the economic process of transformation of natural resources cannot alter the sum of matter-energy, merely its form. In a sense, then, resources are not so much depleted as changed. For example, there is as much copper or iron within the total system as there was thousands of years ago. It does not disappear, but through the operation of the second law of thermodynamics, it is less available.

This law states that the general direction of matter-energy is from low-entropy (ordered and concentrated) states to high-entropy (disordered and dissipated) states. It is these laws that determine that, for example, there will always be more biomass of plants on the Earth than herbivores, and more biomass of herbivores than carnivores. Human economic systems are very much less efficient than nature's, however, and the amount of available and usable matter-energy is constantly being degraded to low-availability materials and low-energy heat. Neither of these 'waste products' can be recovered without further use of energy, which itself is degraded in the process. However, designing a cyclically based and resource-energy-efficient economy would greatly reduce the total entropy effect. At present the massive global distribution of materials, pollution and waste – both intentionally and unintentionally – seems designed to maximize entropy.

Indeed, pollution and waste may be defined as a high-entropy state. Globally, we are now in the situation where the benefit of economic activity is being countered by the negative impact of pollution on the productivity of natural systems and their capacity to deal with pollution. Both locally and internationally, we tend to deal with this problem by attempting to ensure that the pollution occurs 'elsewhere', perhaps at the local landfill site, or in a developing country where emission laws are less stringent than our own. There is, however, no escape from entropic laws and the increasing cost of environmental clean-up, and environmental damage (such as acid rain, soil loss, loss of biological diversity, increased ultraviolet radiation through ozone depletion and so on) leading to loss of productivity and the threat of the deterioration of life-support systems is beginning to bring this point home forcibly.

Further, the entropy laws predict and govern the phenomenon of global warming. In 1970, Ehrlich warned that they 'pose the threat that man may make this planet uncomfortably warm with degraded energy well before he runs out of high-grade energy to consume', a predicament which now is increasingly likely to overshadow fear of running down fossil fuel supplies (Ehrlich 1970: 55). Entropy is also involved in inflation. As materials and energy are dissipated, and new sources become less accessible, proportionately more energy and money have to be spent to win and process them, thus increasing costs.

These principles make suspect the claims that more undifferentiated economic growth is needed to deal with pollution problems or poverty. As Daly has said, economic growth as currently pursued 'may increase costs faster than benefits and initiate an era of uneconomic growth which impoverishes rather than enriches' (quoted in Worldwatch Institute 1987: 6). Industrial societies may have reached the 'entropy state' predicted by Henderson where 'complexity and interdependence have reached the point where the transaction costs that are generated equal or exceed the societies' productive capabilities' (Henderson 1978: 84).

The fundamental dilemma posed by resource demand and environmental capacity is reflected in the Brundtland Report, which anticipated 'a five- to tenfold increase in world industrial output by the time world population stabilizes sometime in the next century' (WCED 1987: 213). This growth has 'serious implications for the future of the world's ecosystems and its natural resource base', but the report's answer is to change the content of growth 'to make it less material and energy intensive and more equitable in its impact' (WCED 1987: 52). Given past and present wasteful use of resources, especially energy, there is a great deal of scope for much more efficient use and this is fast becoming a key thrust of global energy scenario research. Ultimately, however, entropy ensures that increasing production and consumption must involve increasing resource use and pollution (Ekins 1986: 11). Indeed, the report states that 'on balance, environmental problems linked to resource use will intensify in global terms' (WCED 1987: 32).

Some scientists and politicians point to technological solutions to resource issues, through making substitutions, developing biotechnology and so on. Unless the fundamental materialist value base of economic activity changes, however, these innovations may only serve to increase net throughput of resources, whilst introducing possible further threats to the environmental integrity of ecosystems through the introduction of novel pollutants. At best, such innovations might buy time until such value changes, if they occur on sufficient scale, can produce real structural change.

TOWARDS AN ECOLOGY OF RESOURCES

Clearly, a fundamental change is necessary. As Brown says: 'Continuing on a business-as-usual path virtually assures severe economic disruption, social instability, and human suffering' (Brown 1990a). Ultimately, the only sustainable path is one where local and global economies support rather than destroy the natural economy. Indeed, according to Commoner, environmental deterioration is a signal that we have failed, thus far, 'to achieve this essential accommodation' (Commoner 1972). Such are the close links between ecology and economy that any economic strategy is, whether or not intended, an environmental strategy. Therefore, the two have to be considered as one system if both are to survive, and this means moving towards a more equitable and cyclical economy.

As Boulding says: 'if the human race is to survive, it must develop a cyclical economy in which all materials are obtained from the great reservoirs – the air, the soil, the sea – and are returned to them, and which the whole process is powered by solar energy' (Boulding 1978: 296). This conclusion is reached by regarding the Earth as a closed system (except for incoming solar energy).

A unique research project, Biosphere 2, being conducted in the Arizonan desert is as far as possible emulating natural systems in order to sustain human and plant life for two years in a sealed 3.5-acre capsule. Natural systems appear to reverse entropic processes locally, ensuring that all material waste outputs are inputs for another part of the system. 'While 100% recycling is thermodynamically impossible, natural ecosystems come as close as possible to the ideal state of a balanced budget between production and consumption' (Rifkin 1985: 144).

Whilst the Biosphere 2 experiment continues, a more urgent need is to build an eco-economy on Biosphere 1 (the Earth). A low-entropy, low-pollution, resource-conserving economy would be one where:

• Natural resources are regarded as the real and primary wealth, in terms of stock, of their productive renewing ability and capacity to assimilate and recycle society's waste products.

- As far as possible, production is switched from reliance on non-renewable to renewable resources which are not taken at a rate that exceeds the sustainable yield of natural systems.
- Throughput of energy and materials is minimized.
- The production of pollution and waste is minimized, (recognizing that pollution cannot be avoided in the economic process) and is not discharged into the environment at a rate higher than the ability of natural systems to degrade it.
- Account is taken of all environmental costs in extraction, production, consumption and disposal of materials (so that, for example, it is not cheaper to use new raw materials than recycle materials).
- Maximum durability, reuse, repair and recycling of materials (in that order) is encouraged as far as possible ('doing more with less').
- The production of toxic and synthetic waste is minimized and recycled in closed systems.
- Local resources are used to meet local needs.

At the same time, the entropy generated through massive transfer of resources and materials across the globe, involving huge energy costs and often disruption of local ecosystems and economies, will need to be reduced. Ideally, as some writers have observed, a pattern of 'bioregionalism' would be encouraged where healthy and largely self-reliant local economies would be as far as possible built on the potential afforded by sustainable use of local and regional ecosystems.

Finite, non-renewable resources would be spent carefully to buy the time necessary to shift towards a sustainable low-energy path. However, such ideas and criteria are only just beginning to infiltrate economic thinking. For example, at present, the gross national product (GNP), which is the universal indicator of economic performance, takes no account of the economic benefit of healthy ecosystems, or of their degradation. 'Development policies, whose principal goal is to increase GNP, as opposed to meeting basic human needs, can encourage developing countries to liquidate stocks such as forests as rapidly as possible to increase the flow of money through the economy' (Hall 1990: 100).

An increasing number of economists have made and are formulating an eco-economics which can more nearly cost social and environmental costs and benefits. Daly, for example, has advocated the 'steady state' economy where low throughputs, in keeping with the assimilative capacities of the ecosphere, maintain constant but not static stocks of artefacts and people (Ekins 1986: 13). Whilst governments are showing signs of interest in environmental accounting, Hall remarks that 'the influence of such thought to routine economic analysis seems very small and is still inadequate' (Hall 1990: 99).

However, there are some encouraging signs. Increasing costs of energy and resources, and the savings that can often be made from recycling wastes, are resulting in increasing energy efficiency and reuse of materials in many countries, which has in turn slowed their growth in resource consumption. Energy intensity – the ratio of energy to GNP – declined in fourteen European countries, the United States, Canada and Japan between 1970 and 1986 (World Resources Institute and IIED 1988). A World Resources Institute study has shown that the amount of energy per capita used in the developed world could, through conservation and efficiency measures, be halved by 2020 without a fall in living standards.

Indeed, the directions that must be taken to achieve sustainable resource use globally are well known (L. Brown 1990c). Yet, as Barbara Ward pointed out some years ago (Ward, in Eckholm 1982: xi), identifying the remedies is no longer the problem. 'The problems are rooted in the society and the economy – and in the end the political structures.' These structures are beginning to shift under the weight of the growing convergence between economic sense, survival strategy and ethical principles.

ETHICAL ISSUES

Ethical questions surrounding resource use spring from the distorted interrelationships – of people with each other, with future generations and with the Earth itself – that seem to characterize our times. In each case, the unsustainability of present resource use patterns intensifies the ethical questions.

Of these questions, which are all related, perhaps the most pressing is the international debt situation which forces less developed countries, and poor communities within them, to destroy natural resource bases in order to earn currency with which to pay interest on loans. Linked to this is the question of equity in sharing global resources, where the rich nations have much greater purchasing and economic power and a determination to maintain accustomed living standards. At the same time, less developed countries are seeking to emulate developed countries and are seeking to raise consumption levels.

Meanwhile, the richer nations, concerned at global environmental issues, are asking the poorer nations to build in environmental protection into economic development plans. For example, India and China are being asked to phase out CFCs in refrigerator production. Their response has been to ask the richer nations to pay for substitute coolants, as it was these nations that damaged the ozone layer in the first place (P. Brown 1990). This exemplifies the broader question of the responsibility of the developed nations in paying for environmental rehabilitation and sound development.

As has become clear, much of the wealth of the developed countries, in the west at least, has been achieved at the cost of Third World environments, and it is here that some of that wealth must be reinvested in planting trees, conserving soil, increasing energy efficiency, protecting biological diversity, and population programmes. The last point, regarding population stabilization, raises an even more difficult ethical problem; yet unless this issue is tackled, malnutrition and environmental degradation may well exact a toll greater than the 14 million children under the age of 5 who already die every year in the developing world (Timberlake and Thomas 1990).

The prospect of achieving a sustainable global economy, however likely or unlikely, raises further ethical questions. While economic growth holds out the prospect – if not the reality – of increasing wealth and standards of living both within and between nations, the problem of equity is contained by the promise of wealth in the future. However, both at local and international level, either widespread recognition that we presently have an unsustainable economy or the attainment of a sustainable economy in steady state or low-growth state would justifiably accentuate already growing demands for a much more even 'slice of the cake'. Unless the crucial issue of equity can be resolved in a move towards a more just international order, 'resource wars' of an economic or military nature may prevail.

The resolution of these issues centres on the possibility of sharing world resources between hemispheres, countries and peoples now living without damaging the ecosphere. Whilst this in itself is an immense challenge requiring fundamental attitudinal and structural change, the further consideration of future generations as yet unborn adds another major dimension to the equation. A diminishing and degraded stock of natural resources passes economic costs down to be paid by our children or generations beyond, but they of course have no say in current decisions which will affect them.

Whilst all these issues are of the utmost concern, some writers argue that they are all anthropocentric in orientation and that this is the root problem from which all others proceed. Indeed, some 'deep ecologists' and 'Gaianists' criticize even the 'spaceship Earth' concept for being a projection of a mechanistic, dualistic and human-centred world view. We need instead to view ourselves as an integral part of the greater reality of the Earth's systems as a living whole, they argue. From this perspective, the reversal of the evolutionary process – the reduction of biodiversity – might be seen as the key ethical issue.

Whether a human-centred or biocentric view prevails, the urgency and interrelatedness of current ethical problems indicate that a wholesale, rather than a piecemeal, reorientation of global society is needed if an equitable and sustainable future is to be assured. It may be that the imperative nature of this goal will induce the necessary change.

TRANSITION

Many observers agree that global society is in a transitory state towards some state of equilibrium between people and the environment, demand and resources, which either will be imposed through catastrophic breakdown and dislocation, or is arrived at through intentioned rapid progress towards a just and sustainable society. Education at all levels, which centres on the ethical and practical implications of sustainability, and takes an holistic rather than a reductionist view of the interrelationship between all areas of human concern and natural systems, is clearly vital. However, the World Conservation Union notes that internationally 'sustainable development hardly enters into the curriculum' (1989: 90). Ideally, education would be expanded massively, as advocated by the Brundtland Report, as a vital resource that could perhaps buy time to address critical natural resource issues. As Myers says: 'If we are to get to grips with our population and environmental problems, we need to appraise the one resource that is in super short supply, time' (Myers 1990: 5).

References

Adams, W. M. (1988) 'Rural protest, land policy and the planning process on the Bakalori Project, Nigeria', *Africa* 58.

Agassi, M., Benyamini, Y., Morin, J., Marish, S. and Henkin, A. (1986) *The Israeli Concept for Runoff and Erosion Control in Semi-Arid and Arid Zones in the Mediterranean Basin*, Emer-Hefer, Israel: Soil Erosion Research Station.

Alkamper, J., Hauffer, W., Matter, H. E., Weise, O. R. and Weiter, M. (1979) 'Erosion control and afforestation in Haraz, Yemen Arab Republic', *Gussener Beitrage Zur Entwicklungsforschung*, Reihe 2.

Ames, B. (1983) 'Dietary carcinogens and anti-carcinogens', *Science* 221.

Anderson (1987) 'A coal gasification combined cycle power plant', *Ambio* 16.

Bailey, M. J. (1980) *Reducing Risks to Life: Measurement of the Benefits*, Washington, DC: American Enterprise Institute.

Baird, J. C. (1986) 'Prediction of environmental risk over very long time scales', *Journal of Environmental Psychology* 6: 233–44.

Baron, J. (1988) *Thinking and Deciding*, Cambridge: Cambridge University Press.

Biswas, A. K. (1980) 'The Nile and its environment', *Water Supply and Management* 4.

—— (1990) 'Conservation and management of water resources', in A. S. Goudie (ed.) *Techniques for Desert Reclamation*, Oxford: Blackwell.

Biswas, A. K. and Arar, A. (1988) *Treatment and Reuse of Wastewater*, London: Butterworths.

Bokhari, S. M. H. (1980) 'Case study on waterlogging and salinity problems in Pakistan', *Water Supply and Management* 4.

Boulding, K. (1966) 'The economics of the coming spaceship Earth', in J. Barr (1971) *The Environmental Handbook*, London: Ballantine.

—— (1978) *Ecodynamics – a New Theory of Societal Evolution*, Beverly Hills, CA: Sage.

Brown, L. (1990a) 'A global action plan', *People* 17 (1): London: International Planned Parenthood Federation.

—— (1990b) 'The illusion of progress', in Worldwatch Institute *State of the World 1990*, New York: Norton; London: Unwin-Hyman.

—— (1990c) 'Picturing a sustainable society', in Worldwatch Institute *State of the World 1990*, New York: Norton; London: Unwin-Hyman.

Brown, L. and Postel, S. (1987) 'Thresholds of change', in Worldwatch Institute *State of the World 1987*, New York: Norton.

Brown, P. (1990) 'India insists on recipe for green fridges', *Guardian*, 29 June.

Carruthers, I. (1981) 'Neglect of O and M in irrigation: the need for new sources and forms of support', *Water Supply and Management* 5.

—— (1985) 'Protecting irrigation investment: the drainage factor', *Ceres* 18.

Chance, M. R. A. (ed.) (1988) *Social Fabrics of the Mind*, Hillsdale, NJ: Erlbaum.

Chandler, W. D. (1986) Worldwatch Paper 72, Washington, DC: Worldwatch Institute.

Chapman, L. J. and Chapman, J. P. (1967) 'Genesis of popular but erroneous psychodiagnostic observations', *Journal of Abnormal Psychology* 6: 193–204.

Chapman, R. J. K. (ed.) (1983) *The Future*, Launceston, Tas: University of Tasmania.

Christian Science Monitor (1988) *Agenda 2000*, 25 July.

—— (1990) 'Hydrogen eyed as fuel for the future', *International CSM*, January 1990.

Clark, S. L. R. (1984) *The Moral Status of Animals*, Oxford: Oxford University Press.

Clegg, G. and Horsman, P. (1990) 'Inputs of organic and inorganic discharges from rivers and direct discharges from the UK', in Marine Forum for Environmental Issues *North Sea Report*.

Club of Rome (1972) *Limits to Growth*.

Cole, D. C. H. and Lewis, J. G. (1960) 'Progress report on the effect of soluble salts on stability of compacted soils', *Proceedings, International Conference on Soil Mechanics and Foundation Engineering*.

Commoner, B. (1972) *The Closing Circle*, New York: Bantam.

Cooke, R. V. and Doornkamp, J. C. (1990) *Geomorphology in Environmental Management*, Oxford: Clarendon Press.

Cooper, C. (1989) 'Taking the Greens seriously', London: Libertarian Alliance.

Council for Environmental Education (CEE) (1990) *Environmental Education and Environmental Policy*, submission on the then proposed government white paper on the environment, Reading: CEE.

Cousteau, J. Y. (1981) 'An inventory of life on our water planet', in M. Richards *The Cousteau Almanac*, New York: Columbus Books/Doubleday.

Department of Education and Science (DES) (1989) *Environmental Education from 5–16*, Curriculum Matters 13, London: HMSO.

Department of the Environment (DOE) (1986) *Assessment of Best Practicable Environmental Options (BPEOs) for Management of Low- and Intermediate-Level Solid Radioactive Wastes*, London: HMSO.

Devall, W. and Sessions, G. (1985) *Deep Ecology: Living as if Nature Mattered*, New York: Gibbs Smith.

Diamant, B. Z. (1980) 'Environmental repercussions of irrigation development in hot climates', *Environmental Conservation* 7.

Dixon, T. (1990) An investigation by von Franeker (1985), quoted by Dixon (director Marine Litter Research Programme, Tidy Britain Group), in Marine Forum for Environmental Issues, *North Sea Report*.

Doxey, G. V. (1975) 'A causation theory of visitor–resident irritants: methodology and research inferences', *Proceedings of the Travel Research Association Sixth Annual Conference*, San Diego, CA.

Drenge, H. E. 1983). *Desertification of Arid Lands*, Chur: Harwood Academic.

Dudley, N. Barret, M. and Baldock, D. (1985) *The Acid Rain Controversy*, London: Earth Resources Research.

Dutton, R. W. (1988) 'The scientific results of the Royal Geographical Society Oman Wahiba Sands Project 1985–1987', *Journal of Oman Studies*, special report no. 3.

Earthlife (1986) *Paradise Lost?*, London: Earthlife Foundation in association with the *Observer*.

Eckholm, E. (1982) *Down to Earth: Environment and Human Needs*, London: Pluto Press.
Ecologist (1990) 20, 5 (September–October).
—— (1991) 21, 2 (March–April).
Ehrlich, P. (1970) *Population, Resources, Environment*, San Francisco: W. H. Freeman.
Eisler, R. (1987) *The Chalice and the Blade: Our History, Our Future*, San Francisco: Harper & Row.
Ekins, P. (ed.) (1986) *The Living Economy*, London: Routledge & Kegan Paul.
Elliott, R. and Gair, A. (eds) (1982) *Environmental Philosophy: A Collection of Readings*, St Lucia, Qu.: University of Queensland Press.
Encyclopaedia of Philosophy (1967) ed. Paul Edwards, New York: Macmillan/ Free Press.
Engel, J. R. and Engel, J. G. (eds) (1990) *Ethics of Environment and Development: Global Challenge and International Response*, London: Bellhaven Press
Environment Canada (1982) *Downwind: The Acid Rain Story*, Ottawa: Ministry of Supply and Services.
Erikson, E. H. (1963) *Childhood and Society*, New York: Norton.
Evans, J. St B. T. (1989) *Bias in Human Reasoning: Causes and Consequences*, Hove, London and Hillsdale, NJ: Erlbaum.
Evernden, L. L. N. (1985) *The Natural Alien: Humankind and Environment*, Toronto: University of Toronto Press.
Ferré, F. (1988) *Philosophy of Technology*, Englewood Cliffs, NJ: Prentice-Hall.
Fischoff, B., Slovic, P. and Lichtenstein, S. (1978) 'Fault trees: sensitivity of estimated failure probabilities to problem representation', *Journal of Experimental Psychology: Human Perception and Performance* 4: 330–4.
Fishkin, J. S. (1982) *The Limits of Obligation*, New Haven, CT: Yale University Press.
Floret, C. and Le Floch, E. (1984) 'Agriculture and desertification in arid zones of northern Africa', UNEP/USSR Workshop on Impact of Agricultural Practices, Batavia, USSR, mimeo.
Fookes, P. G. and Collis, L. (1975) 'Problems in the Middle East', *Concrete* 9.
Fookes, P. G. and French, W. J. (1977) 'Soluble salt damage to surfaced roads in the Middle East', *Journal of Institute of Highway Engineers* 24.
Food and Agriculture Organization of the United Nations (FAO) (1965) *Soil Erosion by Water – Some Measures for its Control on Cultivated Lands*, Agricultural Development Paper 81, Rome: FAO.
Gazdar, M. N. (1987) *Environmental Crisis in Pakistan*, Kuala Lumpur: The Open Press.
Georgescu-Roegen (1974) *The Entropy Law and the Economic Process*, Cambridge, MA: Harvard University Press.
Gerlach, S. A. (1988) 'Nutrients: an overview', in P. J. Newman and A. R. Agg (eds) *Environmental Protection of the North Sea*, Oxford: Heinemann.
Gilette, C. P. and Krier, J. E. (1990) 'Risks, and agencies', *University of Pennsylvania Law Review* 138.
Glover, J. (1984) *What Sort of People Should There Be?*, London: Penguin.
Goldsmith, E. (1985) 'The World Bank: global financing of impoverishment and famine', *Ecologist* 15 (1/2).
Goodpaster, K. E. and Sayre, K. M. (eds) (1979) *Ethics and Problems of the 21st Century*, Notre Dame: University of Notre Dame Press.
Goudie, A. S. (1984) *The Nature of the Environment*, Oxford: Blackwell.
—— (ed.) (1990) *Techniques for Desert Reclamation*, Chichester: Wiley.
Grainger, A. (1990) *The Threatening Desert: Controlling Desertifictaion*, London: Earthscan.

Greenberg, M. R., Sachsman, D. B., Sandman, P. M. and Salomone, K. L. (1989) 'Network evening news coverage of environmental risk', *Risk Analysis* 9: 119–26.

Gribbin, J. and Kelly, M. (1989) *Winds of Change*, Sevenoaks: Headway/ Hodder & Stoughton.

Hahlweg, K. and Hooker, C. A. (eds) (1988) *Issues in Evolutionary Epistemology*, New York: State University of New York Press.

Hall, C. (1990) 'Sanctioning resource depletion: economic development and neo-economics', *Ecologist*, 20 (3).

Hanson, P. P. (ed.) (1986) *Environmental Ethics: Philosophical and Policy Perspectives*, Burnaby, BC: Simon Frazer University.

Heathcote, R. L. (1987) *The Arid Lands: Their Use and Abuse*, London: Longman.

Heidegger, M. (1962) *Being and Time*, Oxford: Blackwell.

Henderson, H. (1978) *Creating Alternative Futures*, New York: Perigree Books.

Herrigel, E. (1985) *Zen in the Art of Archery*, London: Arkana.

Hjulstrom, F. (1939) 'Transportation of detritus by moving water', in P. D. Trask (ed.) *Recent Marine Sediments*, American Association of Petroleum Geologists.

Hobbes, T. (1960) *Leviathan*, Oxford: Blackwell.

Holliday, F. G. T. (Chairman) (1984) *Report of the Independent Review of Disposal of Radioactive Waste in the Northeast Atlantic*, London: HMSO.

Holroyd, M. (1989) *Bernard Shaw, Vol. II, 1898–1918: the Pursuit of Power*, London: Chatto & Windus.

Hooker, C. A. (1982) 'A sheep in wolf's clothing: a critical examination of John Passmore's approach to environmental problems', in R. Elliot and A. Gair (eds) *Environmental Philosophy: A Collection of Readings*, St Lucia, Qu.: University of Queensland Press.

—— (1983) 'The future must be a fantasy', in R. J. K. Chapman (ed.) *The Future*, Launceston, Tas.: University of Tasmania.

—— (1984a) 'Design or perish', *Transactions of the Institution of Engineers, Australia* (August): 6–14.

—— (1984b) 'Benefit–cost–risk analysis in decision making: the normative assumptions', in J. Nicholas (ed.) *Moral Priorities in Medical Research*, Toronto: Hannah Institute for the History of Medicine.

—— (1987) *A Realistic Theory of Science*, New York: State University of New York Press.

—— (1989) 'Towards a philosophy and practice of energy policy making', *Energy Studies Review* 1(2): 130–42.

—— (1990a) 'Toward a categorization of ethical environmental responsibility', unpublished.

—— (1990b) 'Reasoned moral sympathy for all: an analysis of the normative issues underlying health care resource allocation', unpublished.

Hooker, C. A. and Van Hulst, R. (1979) 'The meaning of environmental problems for public political institutions', in W. Leiss (ed.) *Ecology versus Politics in Canada*, Toronto: University of Toronto Press.

—— and —— (1980) 'Institutionalizing a high quality conserver society', *Alternatives* (journal of Friends of the Earth, Canada) 9 (winter): 25–36.

Hooker, C. A., McDonald, R. M., Van Hulst, R. and Victor, P. (1980) *Energy and the Quality of Life*, Toronto: University of Toronto Press.

Huber, P. (1983) 'Exorcists vs gatekeepers in risk regulation', *Regulation*.

—— (1985) 'Safety and the second best: the hazards of public risk management in the courts', *Columbia Law Review* 85.

Hunziker, Z. (1961) Social tourism: its nature and problems, Geneva.

Hurst, P. (1990) *Rain Forest Politics: Ecological Destruction in SE Asia*, London: Zed Books.

Huxley, A. (1958) *The Perennial Philosophy*, London: Fontana.

Ibrahim, F. N. (1982) 'The ecological problems of irrigated cultivation in Egypt', in H. G. Mensching (ed.) *Problems of the Management of Irrigated Land in Areas of Traditional and Modern Cultivation*, Hamburg: International Geophysical Union.

ICI (1990) 'CFCs – the search for alternatives', *Steam* (ICI teachers' magazine) 12 (January).

International Union for the Conservation of Nature and Natural Resources (IUCN) (1980) *The World Conservation Strategy*, Paris: IUCN.

Irvine, S. and Ponton, A. (1988) *A Green Manifesto*, London: Macdonald.

Janis, I. L. (1983) *Groupthink: Psychological Studies of Policy Decisions and Fiascos*, Boston, MA: Houghton Mifflin.

Johnson-Laird, P. N. (1983) *Mental Models*, Cambridge: Cambridge University Press.

Jones, B. (1938) 'Desiccation and the west African colonies', *Geographical Journal 91*.

Kavka, G. (1981) 'The futurity problem', in E. Partridge (ed.) *Responsibility to Future Generations*, New York: Prometheus.

Kennedy, D. (1973) 'Legal formality', *Journal of Legal Studies 2*.

Kolawole, A. (1987) 'Environmental change and the South Chad irrigation project (Nigeria)', *Journal of Arid Environments 13*.

Kovda, V. A. (1977) 'Arid land irrigation and soil fertility: problems of salinity, alkalinity, compaction', in E. B. Worthington (ed.) *Arid Land Irrigation in Developing Countries*, Oxford: Pergamon.

Krippendorf, J. (1987) *The Holiday Makers*, London: Heinemann.

Lal, R. (1987) *Tropical Ecology and Plant Physical Edaphology*, Chichester: Wiley.

—— (1990) 'Water erosion and conservation', in A. S. Goudie (ed.) *Techniques for Desert Reclamation*, Chichester: Wiley.

Leatherhead, T. M. (1990) 'Concentrations of trace organic contaminants in the North Sea', in Marine Forum for Environmental Issues *North Sea Report*.

Le Houérou, H. N. (1977) 'The nature and causes of desertification', in M. H. Glantz (ed.) *Desertification*, Boulder, CO: Westview Press.

Leiss, W. (ed.) (1979) *Ecology versus Politics in Canada*, Toronto: University of Toronto Press.

Leopold, A. (1949) *A Sand County Almanac*, New York: Oxford University Press.

Lord, C. G., Ross, L. and Lepper, M. R. (1979) 'Biased assimilation and attitude polarization: the effects of prior theories on subsequently considered evidence', *Journal of Personality and Social Psychology* 37: 2098–109.

Lovelock, J. E. (1979) *Gaia: A New Look at Life on Earth*, Oxford: Oxford University Press.

Lowin, A. (1967) 'Approach and avoidance: alternative models of selective exposure to information', *Journal of Personality and Social Psychology* 6: 1–9.

Mabbutt, J. A. (1984) 'A new global asssessment of the status and trends of desertification', *Environmental Conservation* 11: 100–13.

McGuire, W. J. (1960) 'A syllogistic analysis of cognitive relationships', in M. J. Rosenbert, C. I. Hovland, W. J. McGuire, R. P. Abelson and J. W. Brehm (eds) *Attitude Organization and Change*, New Haven, CT: Yale University Press.

McKibben, B. (1990) *The End of Nature*, London: Viking.

Mannison, D., McRobbie, M. and Routley, R. (eds) (1980) *Environmental Philosophy*, Canberra: Australian National University.
Marine Forum for Environmental Issues (1990) *North Sea Report*.
Mathieson, A. and Wall, G. (1983) *Tourism: Economic, Physical and Social Impacts*, London: Longman.
Mears, R. (1990) 'Responsibility of travel', *World Magazine*: 33, 86–7.
Merleau-Ponty, M. (1962) *Phenomenology of Perception*, London: Routledge & Kegan Paul.
Miller, P. (1967) *Nature's Notion*, Cambridge, MA: Harvard University Press.
Molina, M. J. and Rowland, F. S. (1974) 'Stratospheric sink for chlorofluoromethanes: chlorine atomic-atalysed destruction of ozone', *Nature*, June.
Moore, G. E. (1903) *Principia Ethica*, Cambridge: Cambridge University Press.
Mortimore, M. (1987) 'Shifting sands and human sorrow: social response to drought and desertification', *Desertification Control Bulletin* 14.
Myers, D. G. and Bishop, G. D. (1970) 'Discussion effects on racial attitudes', *Science* 169: 778–9.
Myers, N. (ed.) (1984) *Gaia: An Atlas of Planet Management*, Garden City, NY: Doubleday/Anchor.
—— (1990) 'People and environment: the watershed decade', *People* 17 (1): London: International Planned Parenthood Federation.
National Academy of Sciences (1983) *Risk Assessment in the Federal Government: Managing the Process*, Washington, DC: National Academy Press.
National Curriculum Council (NCC) (1990) *The Whole Curriculum*, Curriculum Guidance 3, London: NCC.
National Research Council (1980) *Research Priorities in Tropical Biology*, Washington, DC: National Academy of Sciences.
Nelissen, P. H. M. and Stefels, J. (1988) 'Eutrophication in the North Sea', *Rapport* 4 Netherlands Institut voor Onderzoek der Zee.
Nemeth, C. J. (1986) 'Differential contributions of majority and minority influence', *Psychological Review* 93: 23–32.
Nicholas, J. (ed.) (1984) *Moral Priorities in Medical Research*, Toronto: Hannah Institute for the History of Medicine.
Nielsen, K. (1984) 'Global justice, capitalism and the Third World', *Journal of Applied Philosophy* 1: 2.
Nir, D. (1974) *The Semi-Arid World: Man on the Fringe of the Desert*, London: Longman.
NIREX (1987) *The Way Forward: A Discussion Document*, Didcot: NIREX.
—— (1989) *Going Forward*, Didcot: NIREX.
Nisbett, R. E. and Ross, L. (1980) *Human Inference: Strategies and Shortcomings of Social Judgement*, Englewood Cliffs, NJ: Prentice-Hall.
Nitecki, M. H. (ed.) (1989) *Evolutionary Progress*, Chicago: Chicago University Press.
Noble, J. C. and Tongway, D. J. (1988) 'Herbivores in arid and semi-arid range lands', in J. S. Russell and R. F. Isbell (eds) *Australian Soils: The Human Impact*, St Lucia, Qu.: University of Queensland Press.
Nygren, A. (1953) *Agape and Eros*, London: SPCK.
Oakhill, J., Johnson-Laird, P. N. and Garnham, A. (1989) 'Believability and syllogistic reasoning', *Cognition* 31: 117–40.
Observer (1990) 'Death in the air, sickness in the streets', report of conversation with Lubos Benniak, journalist of the *Mlady Suet* newspaper, in Prague, 28 January.
O'Neill, O. (1986) *Faces of Hunger*, London: Allen & Unwin.
Oslo Commission (1989) 'Review on sewage sludge disposal at sea', in *Thirteenth Annual Report*, London: Oslo Commission.

Parfit, D. (1984) *Reasons and Persons*, Oxford: Clarendon Press.

Parker, L. B. and Trumbule, R. E. (1987) *Mitigating Acid Rain with Technology*, Washington, DC: Library of Congress.

Partridge, E. (ed.) (1981) *Responsibility to Future Generations*, New York: Prometheus.

Passmore, J. (1980) *Man's Responsibility for Nature*, London: Duckworth.

Paylore, P. and Greenwell, J. R. (1980) 'Fools rush in: pinpointing the arid zones', *Arid Lands Newsletter* 10.

Pearce, D. W., Markandya, A. and Barbier, E. B. (1989) *Blueprint for a Green Economy*, London: Earthscan; prepared as a report for the Department of the Environment under the title *Sustainable Development, Resource Accounting and Project Appraisal: State of the Art Review*.

Pederson, J. R. (1988) 'Chlorinated environmental pollutants emitted from ocean incineration stocks', *Studsvik Arbetsrapport*, Technical Note Rep. No. EP 88/16, Sweden.

Petty, R. E., Cacioppo, J. T. and Goldman, R. (1981) 'Personal involvement as a determinant of argument-based persuasion', *Journal of Personality and Social Psychology* 41: 847–55.

Rachels, J. (1986) *The End of Life*, Oxford: Oxford University Press.

Rapp, A. (1974) *A Review of Desertification in Africa – Water, Vegetation and Man*, Stockholm: Secretariat for International Ecology.

—— (1987) 'Reflections on desertification 1977–1987: problems and prospects', *Desertification Control Bulletin* 15.

Raven, P. H. (1988) 'Our diminishing tropical rain forests'. in E. O. Wilson (ed.) *Biodiversity*, Washington, DC: National Academy Press.

Rawls, J. (1972) *A Theory of Justice*, New York: Oxford University Press.

Regan, T. and Singer, P. (eds), (1976) *Animal Rights and Human Obligations*, Englewood Cliffs, NJ: Prentice-Hall.

Regens and Rycroft (1988) *Acid Rain Controversy*, Pittsburgh, PA: University of Pittsburgh Press.

Renner, G. T. (1926) 'A famine zone in Africa: the Sudan', *Geographical Review* 16.

Rifkin, J. (1985) *Entropy – a New World View*, London: Paladin.

Ross, W. D. (1930) *The Right and the Good*, Oxford: Oxford University Press.

Routley, V. and Routley, R. (1979) 'Against the inevitability of human chauvinism', in K. E. Goodpaster and K. M. Sayre (eds) *Ethics and Problems of the 21st Century*, Notre Dame: University of Notre Dame Press.

—— and —— (1980) 'Human chauvinism and environmental ethics', in D. Mannison, M. McRobbie and R. Routley (eds) *Environmental Philosophy*, Canberra: Australian National University.

—— and —— (1981) 'Nuclear energy and obligations to the future', in E. Partridge (ed.) *Responsibility to Future Generations*, New York: Prometheus.

Ruckelshaus, W. (1985) 'Risk, science and democracy', *Issues in Science and Technology* 3.

Saad, M. A. H. (1982) 'Distribution of nutrient salts in the lower reaches of the Tigris and Euphrates, Iraq', *Water Supply and Management* 6.

Schumacher, E. F. (1975) *Small Is Beautiful*, London: Abacus.

—— (1977) *A Guide for the Perplexed*, New York: Harper & Row.

Schweitzer, A. (1949) *Civilization and Ethics*, London: A. & C. Black.

Schwing, R. and Albers, W. (eds) (1980) *Societal Risk Assessment: How Safe Is Safe Enough?*, New York: Plenum.

Shalash, S. (1983) 'Degradation of the River Nile', *Water Power and Dam Construction* 35.

Shiva, V. (1987) 'Forests, myths and the World Bank: a critical review of *Tropical Forests: A Call for Action*', *Ecologist* 17 (4/5).

Shiva, V. and Bandyopadhyay, J. (1989) *Ecologist* 19 (3).

Sidgwick, H. (1877) *Methods of Ethics*, London: Macmillan.

Simmons, D. A. (1988) *Environmental Ethics: A Selected Bibliography for the Environmental Professional*, Chicago: Council of Planning Librarians.

Singer, P. (1975) *Animal Liberation*, New York: New York Review.

—— (1979) *Practical Ethics*, Cambridge: Cambridge University Press.

—— (1981) *The Expanding Circle: Ethics and Sociology*, New York: Farrar, Strauss & Giroux.

Singh, D. V., Pal, B. and Kishore, R. (1981) 'Quality of underground irrigation waters in semi-arid tract of District Arga III – Tehsils Bah and Fatehabad', *Annals of Arid Zone* 20.

Slovic, P., Fischhoff, B. and Lichtenstein, S. (1980) 'Facts and fears: understanding perceived risk', in R. Schwing and W. Albers (eds) *Societal Risk Assessment: How Safe Is Safe Enough?*, New York: Plenum.

——, —— and —— (1982) 'Facts versus fears: understanding perceived risk', in D. Kahneman, P. Slovic and A. Tversky (eds) *Judgement under Uncertainty: Heuristics and Biases*, New York: Cambridge University Press.

Smalley, I. J. (1970) 'Cohesion of soil particles and the intrinsic resistance of simple soil systems to wind erosion', *Journal of Soil Science* 21.

Smart, J. J. C. and Williams, B. (1973) *Utilitarianism: For and Against*, Cambridge: Cambridge University Press.

Solow, R. M. (1974) *American Economic Review*, May.

Speth (1989) 'A Luddite recants', *Amicus Journal*, Spring.

Sprigge, T. L. S. (1984) 'Non-human rights: an idealist perspective', *Inquiry* 27: 439–61.

Starr, C. (1969) 'Societal benefit vs technological risk', *Science* 165.

—— (1972) 'Benefit–cost studies in sociotechnical systems', in *Perspectives on Risk–Benefit Decision-Making*, New York: National Academy of Engineering.

Stebbing, E. P. (1935) 'The encroaching Sahara: the threat to the west African colonies', *Geographical Journal* 85.

Stowe, T. J. and Underwood, L. A. (1984) 'Oil spillages affecting sea birds in the UK 1966–1983', *Marine Pollution Bulletin* 15.

Strawson, P. F. (1974) *Freedom and Resentment and Other Essays*, London: Methuen.

Szablocs, I. (1985) 'Salt affected soils – a world problem', in *International Symposium on the Reclamation of Salt Affected Soils*.

Thoreau, H. (1886) *Walden*, London: Walter Scott.

Timberlake, L. and Thomas, L. (1990) *When the Bough Breaks*, London: Earthscan.

Tobias, M. (ed.) (1985) *Deep Ecology*, San Marcos, CA: Avant.

Todd, D. K. (1964) 'Groundwater', in V. Chow (ed.) *Handbook of Applied Hydrology*, New York: McGraw-Hill.

—— (1980) *Groundwater Hydrology*, New York: Wiley.

Trainer, T. (1989) *Developed to Death*, London: Green Print.

Turner, L. and Ash, J. (1975) *The Golden Hordes: International Tourism and the Pleasure Periphery*, London: Constable.

Tversky, D. and Kahneman, A. (1973) 'Availability: a heuristic for judging frequency and probability', *Cognitive Psychology* 5: 207–32.

Tweney, R. D. (1985) 'Faraday's discovery of induction: a cognitive approach', in D. Goodring and F. James (eds) *Faraday Rediscovered: Essays on the Life and Work of Michael Faraday*, Macmillan: London.

United Nations Environment Programme (UNEP) (1987) 'The changing atmosphere', *Environment Brief* (UNEP) 1.

—— (1989a) Speech by Mostafa Tolba, World Environment Day, Brussels, *Bulletin of the UN National Committee of UNEP*, July.

—— (1989b) *North American News* 4 (4), August.

UN World Food Conference Report (1974) *Assessment of the World Food Situation: Present and Future*, Rome, 5–16 November 1974.

Vickers, G. (1970) *Freedom in a Rocking Boat*, Harmondsworth: Penguin.

—— (1980) *Responsibility: Its Sources and Limits*, Seaside, CA: Intersystems Publications.

—— (1983) *Human Systems Are Different*, London: Harper & Row.

Vidal, G. (1989) 'Gods and Greens', *Observer*, 27 August.

Viscusi, K. P. (1983) *Risk by Choice: Regulating Health and Safety in the Workplace*, Cambridge, MA: Harvard University Press.

Wallace, T. (1980) 'Agricultural projects and the land in northern Nigeria', *Review of African Political Economy* 17.

—— (1981) 'The Kano River Project: the impact of an irrigation scheme on productivity and welfare', in J. Heyer, P. Roberts and G. Williams (eds), *Rural Development of Tropical Africa*, London: Macmillan.

Ward, B. (1982) 'Foreword' to E. Eckholm *Down to Earth: Environment and Human Needs*, London: Pluto Press.

Wason, P. C. (1960) 'On the failure to eliminate hypotheses in a conceptual task', *Quarterly Journal of Experimental Psychology* 12: 129–40.

Watson, Gill, Stolarski and Hampson (1986) *Present State of Knowledge in the Upper Atmosphere*, NASA Publication 1162.

Wellings, F. M, (1982) 'Viruses in groundwater', *Environment International* 7.

Whyte, W. H. (1956) *The Organization Man*, New York: Simon & Schuster.

Wilbur, K. (1981) *No Boundary: Eastern and Western Approaches to Personal Growth*, Boston, MA: Shambhala.

Wilkins, M. C. (1928) 'The effect of changed material on the ability to do formal syllogistic reasoning', *Archives of Psychology* 16 (102).

Williams, B. (1985) *Ethics and the Limits of Philosophy*, London: Fontana/Collins.

Wordsworth, W. (1971) *A Choice of Wordsworth's Verse*, selected by R. S. Thomas, London: Faber.

World Bank (1990) *World Development Report*.

World Commission on Environment and Development (WCED) (1987) *Our Common Future* (the Brundtland Report), Oxford: Oxford University Press.

World Conservation Union (1989) *From Strategy to Action*, Gland, Switzerland: IUCN.

World Meteorological Organization (WMO) (1986) *United Nations Conference on Climatic Variations and Associated Impacts, Villach, Austria, October 1985*, proceedings, Geneva: WMO.

World Resources Institute (1985) *The Global Possible: Resources, Development and the New Century*, New Haven, CT: Yale University Press.

World Resources Institute and International Institute for Environment and Development (IIED) (1988) *World Resources 1988–89*, New York: Basic Books.

Worldwatch Institute (1987) *State of the World 1987: A Worldwatch Institute Report on Progress toward a Sustainable Society*, New York/London: Norton.

—— (1990a) *Clearing the Air*, Washington, DC: Worldwatch Institute.

—— (1990b) *State of the World 1990: A Worldwatch Institute Report on Progress toward a Sustainable Society*, New York/London: Norton.

Zimmerman, E. (1951) *World Resources and Industries*, New York: Harper & Row.

Index